Modelling and Management of Engineering Processes

Michael Schabacker · Kilian Gericke
Nikoletta Szélig · Sándor Vajna
Editors

Modelling and Management of Engineering Processes

Proceedings of the 3rd International Conference 2013

Springer

Editors
Michael Schabacker
Nikoletta Szélig
Sándor Vajna
Information Technologies in Mechanical Engineering
Otto-von-Guericke University Magdeburg
Magdeburg
Germany

Kilian Gericke
Research Unit in Engineering Science
University of Luxembourg
Luxembourg
Luxembourg

ISBN 978-3-662-44008-7 ISBN 978-3-662-44009-4 (eBook)
DOI 10.1007/978-3-662-44009-4

Library of Congress Control Number: 2014949365

Springer Heidelberg New York Dordrecht London

Printed on acid-free paper

Springer is part of Springer Science+Business Media (www.springer.com)

Preface

Dear friends and colleagues,

as we know, Engineering processes have always been the glue that holds together all activities within product development and design. Engineering processes structure these activities appropriately and secure their reasonable processing. They ensure the correct and timely use of appropriate approaches & procedures, methods, data, and tools in order to improve the design procedures, improve products and services, and properly document both the resulting product as well as its development processes. It has been both the aim of the SIG MMEP (Modelling and Management of Engineering Processes) and of its conferences to contribute to a smart and smooth definition, application, and navigation of Engineering Processes.

I founded the SIG MMEP in 2003 based on discussions at different ICEDs, Rigi meetings, and IPD Workshops that clearly showed the necessity of consolidating the definition, the prospect, and the handling of processes in our Engineering environment. At ICED 07 in Paris I was pleased that John Clarkson agreed to share the SIG leadership with me in order to compare and to put together different approaches of managing Engineering processes and projects in turbulent environments, on which both our institutions have had a long and successful research history. We were lucky that Peter Heisig could be convinced to become a member of the team.

The MMEP conference series were launched in 2010 as a bi-annual event providing an international platform to highlight and to discuss industry best practices alongside leading edge academic research. The second MMEP conference in 2012 focussed on exploring potential synergies between different modelling approaches, and discussed future directions both in managing and researching engineering processes. The participants at MMEP 2012 decided to meet again at ICED 13 in Seoul and for the third MMEP conference in Magdeburg.

In 2013 we celebrated the 10th anniversary of our SIG. On behalf of my co-editors it was our pleasure organising the 3rd International Conference on MMEP 2013 at the Wasserburg (moated castle) of Gommern, close to Magdeburg, where it all started at the Otto-von-Guericke University. We hope that the participants enjoyed the conference. The papers chosen for this proceeding were selected by reference to blind reviews and discussions after their respective presentations undertaken by the participants. These papers represent the areas of process modelling, process optimisation, multi-project and process management, Key Performance Indicators, Lean Product Development and others. We would like to thank all those authors and reviewers who have contributed to the preparation of this book. We also thank Ms. E. Hestermann-Beyerle and Ms. B. Kollmar-Thoni from Springer for the smooth and constructive cooperation.

And, after ten years, it is a good practice to hand over the SIG leadership to younger people. I am very happy that Dr.-Ing. Kilian Gericke, University of Luxemburg, and Prof. Dr. Claudia Eckert, The Open University (UK), agreed to co-chair the SIG MMEP. Having known them well for a long time, I have no doubt that they will continue with the fruitful SIG work, of course (and hopefully!) with other aspects than we used to prefer in our time. But what will surely remain is that this our SIG will keep fascinating and challenging and beneficial to all its members.

Let me conclude. It was a highlight to work with you all! Good bye, and Vivat, Crescat, Floreat to our SIG on Modelling and Management of Engineering Processes!

Prof. Dr.-Ing. Dr. h.c. Sándor Vajna
Otto-von-Guericke University Magdeburg, Germany
July 2014

Contents

Chapter 1

A Process Taxonomy Model for Engineering Design Research

N. Chucholowski, F. Schoettl, W. Bauer, S. A. Schenkl, F. Behncke and U. Lindemann

1.1 Introduction

Research on engineering processes plays a major role within engineering design research, since "the bulk of the effort involved in product development lies in perfecting the underlying processes" (Panchal *et al.*, 2004). Processes are getting even more important when considering the emerging research topic of product-service systems (PSS): PSS are an integration of product and service elements in one market offer (Baines *et al.*, 2007). Thereby not only design processes have to be considered but also the service product itself may be modeled as a process (Bullinger *et al.*, 2003).

There are many possible aspects to consider when doing research on engineering processes. Such process aspects comprise research on e.g. different processes among the product lifecycle (Panchal *et al.*, 2004), different activities on processes (Browning and Ramasesh, 2007), different process knowledge (Hubka and Eder, 1996), different characteristics of processes (Maier and Störrle, 2011) or the modeling of processes (Browning *et al.*, 2006). Research on engineering processes may compromise describing and modeling industrial processes, prescriptively defining processes as well as developing supporting methods and tools. The goals of these activities are amongst others to raise efficiency and effectiveness, resilience, adaptability or transparency of processes in the industrial practice.

Consequently, there is a big number of different research efforts and generated knowledge, which are hard to overlook. There are frameworks that attempt to order design knowledge respectively engineering design research efforts and findings, e.g. (Horváth, 2004). They are supposed to "help researchers to locate their work in the global picture of engineering design, [...], granters to make decisions about the possible fields of investments, and educators to organize subject materials for various design courses" (Horváth, 2004). So far, process research is only a part of

M. Schabacker et al. (eds.), *Modelling and Management of Engineering Processes*, DOI 10.1007/978-3-662-44009-4_1

these frameworks and is addressed on a too rudimentary level to distinguish the above described aspects and the related research on these aspects.

We developed a taxonomy model that enables not only to regard specific types of processes in the product lifecycle and specific activities, which can be applied on these processes, but also the allocation of research efforts and results among these processes, activities and even the "research on research" on a meta-level. Furthermore, the transition of research into practice is addressed, which was identified as a lack of design research taxonomies by Fulcher (1998).

The process taxonomy for engineering processes as an explanatory model allows classifying, mapping and delimitating specific research activities and results. It considers the wholeness of aspects regarding the management and modeling of engineering processes.

1.2 Research Methodology

In order to get an integrated understanding of the topic as a foundation for the taxonomy model for engineering processes, we performed a literature review, considering publications on process research. The following keywords have been applied: design process, design process research, research model design process, engineering process research and process research. Besides that we have indexed 13 engineering process-related research projects funded from public organizations such as the German Research Foundation (DFG) as well as industry funded research projects. Based on that, we have prescriptively set up the taxonomy model considering relevant aspects of engineering processes on the two layers research and industry. For verification, we have discussed the model in a working group consisting of researchers working on different aspects of engineering design processes, such as product planning, engineering change management and production planning. Furthermore, the model was applied on several examples of process research to test its internal validity.

1.3 Process Research in Engineering Design Literature

This section gives a short overview of literature addressing design research in general, whereas the focus is on identified literature that deals with design process research. Afterwards, derived implications are described.

A look on literature about design research and design science reveals that the research on processes plays a major role (e.g. Fulcher, 1998; Horváth, 2004; Hubka and Eder, 1996; Panchal *et al.*, 2004; Pugh, 1990; Ullman, 1992). Most of their research efforts aim to classify and characterize research on design processes. A categorization of the process research space is necessary in order to identify appropriate tools and techniques for process research and to place research efforts in perspective (Ullman, 1992).

Amongst others but profoundly inspired by the technical systems theory (Hubka and Eder, 1996), Horváth (2004) categorized engineering design research in a framework. He considered process knowledge as one category within engineering design research; including the domains *design process*, *artifactual process* and *implicated process* (cf. Figure 1.1). Studying, modeling and the optimization of the design process itself are so called research trajectories within the domain *design process*. *Artifactual processes* address existential, operation, application and service processes of products. Research in the domain of *implicated processes* addresses all processes that are related to the realization and utilization of a product, e.g. technological, production, sales/supply and reclaiming processes (Horváth, 2004).

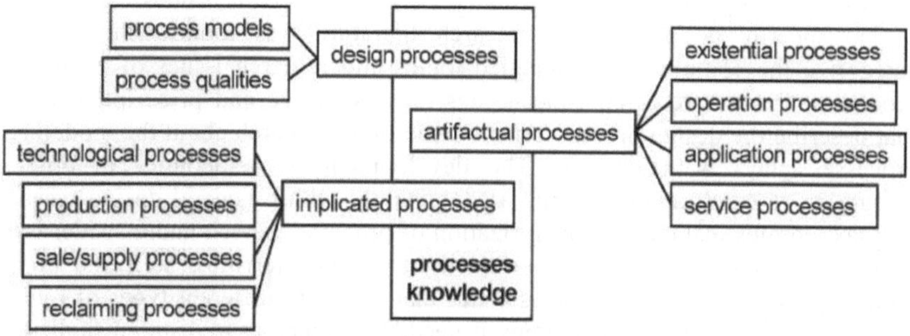

Figure 1.1. The domains and trajectories within the category "process knowledge" in the framework of reasoning by Horváth (2004)

The consideration of artifactual and implicated processes extends the perspective on design process research. Also Panchal *et al.* (2004) proclaim that design process research should integrate perspectives from the whole product lifecycle. When all lifecycle phases of a product are regarded during its design, the design process has to meet certain requirements. Based on this, Panchal *et al.* (2004) list the following research issues:

- modeling design processes;
- computational representations of design processes;
- storage of design information;
- developing metrics for assessing design processes;
- configuring design processes;
- integrated design of products and design processes;
- integrating design processes with other processes in product lifecycle management.

Hence the design process is not the only process that design research has to deal with. Designers have to consider all processes within the product life cycle.

For example, also the service process has to be designed when considering the design of product-service systems. As a lot of different processes have to be considered, it is hard to keep record which process is meant when speaking about engineering processes.

When looking at fundamental design literature (e.g. Hales and Gooch, 2004; Lindemann, 2009; Pahl *et al.*, 2007; Suh, 1990; Ulrich and Eppinger, 2003), chapters about the design process deal with the definition of the process itself, i.e.: What activities, methods, attributes are part of the process and what characteristics describe the process? Also more specific design process research (e.g. Clarkson and Eckert, 2005; Gericke and Blessing, 2011; Maier and Störrle, 2011) predominantly tends to define the process regarding different disciplines or process characteristics. Another often addressed issue in specific design process research is process modeling (cf. the literature review by Browning *et al.*, 2006).

No specific design process research literature was found which aims to categorize the complex and entangled topics and research issues of engineering processes. Process research concentrates on descriptive and prescriptive models that describe the design process. At the same time, research about the modeling of processes plays a major role. The development of taxonomies in order to categorize research areas happens predominantly within general design research. The taxonomies include the categorization of process research and consider issues such as the design processes itself (activities, characteristics, etc.) and process modeling. Additionally it becomes clear, that there are different types of relevant processes to investigate. A taxonomy is needed that enables to distinguish different processes on the one hand and different activities addressing these processes on the other hand.

1.4 A Taxonomy Model to Classify Engineering Design Process Research

In this section we explain the purpose of the model, describe its structure and individual parts as well as intended applications, and illustrate the use of the model.

1.4.1 Motivation and Intended Purpose of the Model

From a process research point of view, there is a need for support to classify research activities. By this, not only a classification in the present landscape of process research and a differentiation from other research projects is possible, but also research gaps can be identified. Potential users of the taxonomy model could be e.g. research institutions, researchers or design educators. Researchers and research institutions can create a profile of their research topics that is easy to understand and they can also identify and visualize future research fields that should be aimed at. For educators in the area of design the model can give a clear

overview on the different aspects that are relevant when dealing with engineering processes.
A significant added value compared to existing models is the level of abstraction of our approach, which allows an integrated and lifecycle-oriented consideration according to Systems Engineering. Specifically, these are the possible connections between research and practice, activities and considered processes in the product lifecycle as well as their interfaces, which is seen as particularly important (Fulcher, 1998).

1.4.2 Description of the Model

The described purposes result in the structure of our taxonomy model, which possesses three dimensions spanning a rectangular space (cf. Figure 1.2). The first axis *lifecycle processes* embodies several types of processes from the lifecycle phases of a system in order to classify all relevant processes used in mechanical design. The second axis *activities* in the basic layer shows major activities that are applied to a process. The third dimension allows to model an additional meta-level which regards research about the processes behind the activities and processes shown in the bottom layer. This level is necessary to be able to also classify e.g. research methodologies in process research. The combination of the three dimensions builds up our taxonomy model with 50 points in the defined space. For the classification of any kind of process-related activities, characteristics and the important interfaces in between, we have extended the points to cubes. On the one hand, that represents the steadiness and consistency of processes beyond lifecycle phases which is frequently the case in reality. On the other hand, the cubes facilitate a precise differentiation of several activities and processes along their edges or in their volume. Moreover research projects or activities with diverse practical relevance can be properly allocated in the model using different positions in the vertical dimension.

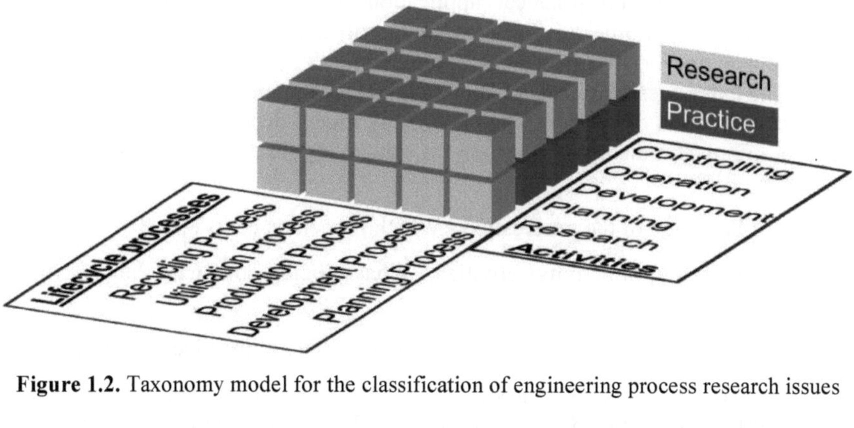

Figure 1.2. Taxonomy model for the classification of engineering process research issues

The processes that occur within the lifecycle of a system in one dimension can be summarized as product planning, development, production, utilization, and recycling process (Haskins and Forsberg, 2011; Hepperle *et al.*, 2009). Products, services or product-service systems can be subject of these processes and lifecycle phases. This axis determines what type of process is considered.

The other dimension is formed by activities that can be applied to the different lifecycle processes. Based on the process taxonomy by Ullman (1992) and the activities applied on processes listed by Browning and Ramasesh (2007), we selected four major activities for the classification: planning, development, operation and controlling. Considering design research, research can be applied to each of the listed system lifecycle processes. Therefore, the activity "research" is added to the second axes. The bottom layer, which is spanned by the two dimensions, maps "Activities on Processes". The process behind the activity "research on a lifecycle process" (e.g. research on development processes) is predominantly performed by academia. For this we chose grey colored cubes. The blue colored area can be interpreted as work with process issues in practice, such as the development of a production process, the planning of a development process or the controlling of a (product) planning process (predominantly performed by practitioners).

The last dimension complements the taxonomy model with research activities that are not represented in the bottom layer. It enables to classify also research, which is done on the activities that can be applied to processes. An example would be the "research on research on development processes". This means how to do research on development processes. Another example is the research on planning production processes, i.e. the "investigation of planning production processes" or how to plan production processes. Hence, the whole second layer and the research activity on the bottom layer incorporate process research in general.

The model also allows referring to interfaces between research and practice. The question of the transferability of research into practice e.g. in terms of controlling planning processes can be located at the transition between the corresponding gray and blue cube downwards. Basically, two different use cases according to the direction of transfer are conceivable. Bringing a scientifically developed method into practical application or transferring problems into a scientific context in order to find a solution with universal scope. Both use cases may include one or two interfaces between practice (blue) and research (grey) areas. The latter possibility is explained later in an example.

The application of the developed taxonomy model depends on the purpose and the user group. But the basic sequence for classifying research efforts or projects is the same:

1. Selection of the process(es) - Which process is considered?
2. Selection of activity/activities - What is being done with the concerning process?
3. Selection of a perspective - Is it about research, practice or a transfer?

After allocating process issues in the model, there are several possibilities to use the results. Since this depends on the specific use case and the perspective, we want to give some exemplary proposals. Researchers may allocate a project and find potential use cases for practical application of their results or for validating their approaches. Furthermore adjacent cubes show potential fields for future research. We strongly recommend to use the taxonomy model in its presented set-up without adding or removing elements, since we attempted to find the lowest common denominator in engineering process research. If there is a need for more detailed investigations, both axis (the lifecycle processes and the activities on processes) can be particularized. For example, the lifecycle process *development process* can be decomposed to more detailed process steps (e.g. concept design, detailed design, etc.) or the activity *controlling* can be split into measuring and adapting to gain a more detailed insight on the object of investigation. However, we recommend to not decompose the lifecycle processes and activities any further in order to keep the simplicity of the model as its strength.

1.4.3 Use Cases

The description and application of our taxonomy model is followed by an evaluation in this subsection. The focus is mainly on the model structure since the basic content of the three axes consists of commonly accepted knowledge in process research. The purposeful adjustment and the proper interplay of the dimensions are demonstrated in four examples, which represent different use cases according to the mentioned purpose.

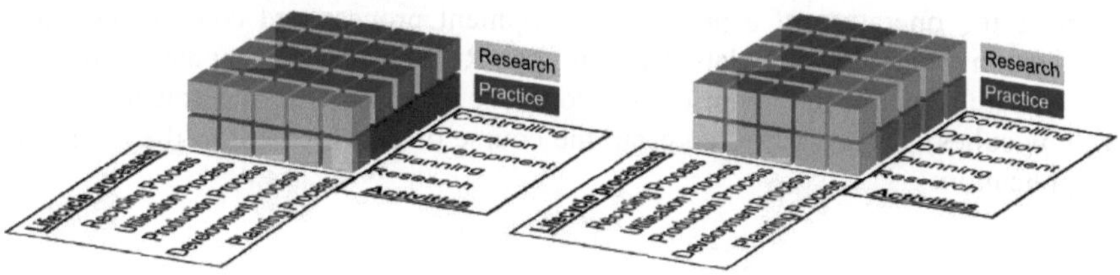

Figure 1.3. Allocation of a resarch methodology (l.) and a research institute (r.)

Figure 1.3 shows the allocation of a research methodology as well as a competence profile of a research institute. On the left side, the established design research methodology (DRM) (Blessing and Chakrabarti, 2009) is allocated to the five yellow cubes in the row of research activity in the upper layer of the model. DRM is research about research in design, with design considering all lifecycle phases of a system (i.e. product). Furthermore, this taxonomy model itself as contribution to process research fits into the same area as DRM because of its universal scope. On the right side, the competences of a research institute e.g. the Institute of Product Development at the Technical University of Munich are illustrated by the colored area. It covers activities like planning, developing, operating and controlling product planning processes or development processes in the context of industrial projects (orange). Research on these mentioned activities, research on the planning process and development process itself and finally research on research activities constitute the scientific part of the institute's competences (yellow).

To emphasize the single cubes and their interfaces, Figure 1.4 shows the allocation of a product development method and its transfer from research into practice. Some cubes were removed in this illustration to ensure a clear view on the relevant elements.

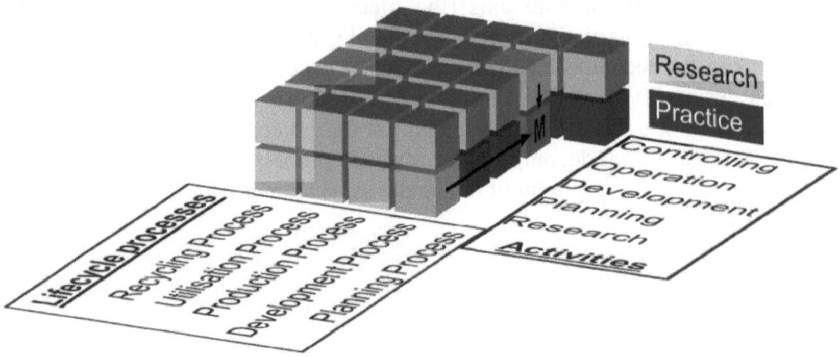

Figure 1.4. Allocation of a development method the transer into practice

The Munich Procedure Model („Münchner Vorgehensmodell") by Lindemann (2009) serves as an example for a product development method, that can be understood as an instruction on the product development process. This method supports the operation of a product development process and consequently it is allocated in the orange cube labeled with "M". Research on development processes and research on the operation of development processes are the scientific basis (yellow cubes). Both arrows represent the course of development to the final field of application.

1.5 Conclusion and Outlook

Taxonomies in design research are used to clarify and to simplify research topics (Fulcher, 1998). The presented process taxonomy model works as a descriptive taxonomy what enables to allocate research efforts in the global field of engineering processes (Horváth, 2004). As Fulcher (1998) stated, the often proposed two-dimensional or hierarchical taxonomies are unlikely to represent the various and complex research topics within engineering design in an adequate depth. By using three dimensions in this process taxonomy model, process research can be classified more precisely by indicating what kind of processes and what activities on the process are considered. Additionally, research on process definitions (characteristics, attributes, etc.), research on activities on processes, and research on research about processes can be differentiated. Besides the application by researchers and educators, the model also enables practitioners to border their competences and service portfolio against competitors as well as take strategic decisions based on the resulting transparency.

As mentioned earlier, specific research about the modeling of processes still draws the research agenda for both industry and academia due to its relevance.

Process models can be seen as tools used in every single cube in the taxonomy model. This implies, that the cubes could be more detailed regarding their content. In practice, there are process models for e.g. the development process as well as for the planning of development processes. Further, the modeling of these processes can be addressed in process research and it is always aligned with process attributes such as tasks, duration or performance indicators etc. Besides process modeling, requirements management, simulation, evaluation and other cross-cutting issues can also be seen as an artifact included in every cube in the taxonomy model.

Another conceivable differentiation of research efforts is to distinguish objects of research and research results. The allocation of a research project to one cube can have two intents: First, the process aligned to this cube is the object of investigation. Second, the results of the project have impact on the cube. For example, one can either look at production processes in order to optimize the product for this process (i.e. the process is object of the investigation) or the research investigates the product in order to develop an ideal production process (i.e. the research results regard the process). The differentiation of these aspects in the taxonomy model is still pending, so that its confirmation represents a first step of future research.

Moreover, the allocation of more examples or use cases is inevitable to ensure that all conceivable aspects of engineering design can be classified within the proposed taxonomy model. Furthermore, there is still a lack of experience where to allocate requirements and results of processes. Interfaces between the cubes have to be described in more detail, due to possible different input-output dependencies between processes and activities. They should also be modeled within the taxonomy model in order to emphasize their important role. Further, the reference to objects which are located in the center of the volume is unsatisfactory because they are hidden by others and therefore limits the usability of the presented taxonomy model. Moreover, the visualization needs further improvement in order to specify the object of the lifecycle processes and distinguish between products, services or PSS.

1.6 References

Baines, T. S., Lightfoot, H. W., Evans, S., Neely, A., Greenough, R., Peppard, J., Roy, R., Shehab, E., Braganza, A., Tiwari, A., Alcock, J. R., Angus, J. P., Bastl, M., Cousens, A., Irving, P., Johnson, M., Kingston, J., Lockett, H., Martinez, V., Michele, P., Tranfield, D., Walton, I. M. and Wilson, H. (2007) 'State-of-the-art in product-service systems.', Proceedings of the Institution of Mechanical Engineers, Part B: Jounal of Engineering Manufacture, vol.10, no.221, pp.1543-1552

Blessing, L. and Chakrabarti, A. (2009) DRM: A Design Reseach Methodology, London, Springer

Browning, T. R., Fricke, E. and Negele, H. (2006) 'Key concepts in modeling product development processes', Systems Engineering, vol.9, no.2, pp.104-128

Browning, T. R. and Ramasesh, R. V. (2007) 'A Survey of Activity Network-Based Process Models for Managing Product Development Projects', Production and Operations Management, vol.16, no.2, pp.217-240

Bullinger, H.-J., Fähnrich, K.-P. and Meiren, T. (2003) 'Service engineering - methodical development of new service products.', International Journal of Production Economics, vol.85, pp.275-287

Clarkson, J. and Eckert, C. (2005) Design process improvement: a review of current practice, London, Springer

Fulcher, A. J. (1998) 'A Taxonomy of Design Research Topics by Multivariate Agglomerative Clustering', Journal of Engineering Design, vol.9, no.4, pp.343-354

Gericke, K. and Blessing, L. (2011) 'Comparisons of design methodologies and process models across disciplines: a literature review', Proceedings of the 18th International Conference on Engineering Design, Design Society

Hales, C. and Gooch, S. (2004) Managing Engineering Design, London, Springer

Haskins, C. and Forsberg, K. (2011) 'Systems Engineering Handbook: A Guide for System Life Cycle Processes and Activities; INCOSE-TP-2003-002-03.2. 1', Incose

Hepperle, C., Thanner, S., Mörtl, M. and Lindemann, U. (2009) 'An integrated product lifecycle model and interrelations in-between the lifecycle phases', 6th International Conference on Product Lifecycle Management, Bath

Horváth, I. (2004) 'A treatise on order in engineering design research', Research in Engineering Design, vol.15, no.3, pp.155-181

Hubka, V. and Eder, W. E. (1996) Design science: introduction to needs, scope and organization of engineering design knowledge, London, Springer

Lindemann, U. (2009) Methodische Entwicklung technischer Produkte: Methoden flexibel und situationsgerecht anwenden, Berlin, Springer

Maier, A. and Störrle, H. (2011) 'What are the characteristics of engineering design processes?', International Conference on Engineering Design (ICED11). Kopenhagen, Denmark

Pahl, G., Beitz, W., Feldhusen, J. and Grote, K.-H. (2007) Engineering design: a systematic approach, London, Springer

Panchal, J. H., Fernández, M. G., Paredis, C. J., Allen, J. K. and Mistree, F. (2004) 'Designing design processes in product lifecycle management: Research issues and strategies', Proceedings of the ASME design engineering technical conference, Salt Lake City, UT, USA

Pugh, S. (1990) 'Engineering Design—Unscrambling the Research Issues', Journal of Engineering Design, vol.1, no.1, pp.65-72

Suh, N. P. (1990) The principles of design, New York, Oxford University Press

Ullman, D. G. (1992) 'A taxonomy for mechanical design', Research in Engineering Design, vol.3, no.3, pp.179-189

Ulrich, K. T. and Eppinger, S. D. (2003) Product design and development, Singapore, Tata McGraw-Hill Education

Chapter 2

A Model for Value in Lean Product Development

G. I. Siyam, K. Gericke, D. C. Wynn and P. J. Clarkson

Lean product development has been developed and deployed in an effort to enhance company operations. Understanding value is the first step to becoming 'Lean'. However, the mere translation of value from its conventional interpretation in lean production as ``something the customer will pay for`` does not equate to an effective value orientation in product development (PD). In order to better understand the theoretical context of PD value in research, as well as the potential application of a value orientation in practice, further study is necessary. This paper aims to broaden the understanding of PD value by discussing by linking roles in value creation and delivery to different contexts and phases of the product life cycle.

2.1 Motivation and Requirements

Product development processes have a critical impact in determining the success of an organisation. This is due to their consumption of approximately 75% of the organisation's resources (Millard, 2001) and their role in materialising the product's specifications (McManus, 2005). Therefore, various tools have been adopted to facilitate management and improvement of product development processes, such as the Design Structure Matrix (DSM) and lean value-oriented approaches. However, because of the complexity of product development and the high levels of uncertainty associated with it, these improvement efforts can be difficult (e.g. Pessoa, 2004). Moreover, available support (e.g. methodologies, methods, tools, recommendations and guidelines) are either very abstract, thus needing adaptation before application in a specific context, or they are very

M. Schabacker et al. (eds.), *Modelling and Management of Engineering Processes*, DOI 10.1007/978-3-662-44009-4_2

detailed and can only be applied in a particular situation (Gericke *et al.*, 2013). This limits their application, or application may not provide the desired results

The success of value-oriented lean approaches at the operational level resulted in a wide range of literature aiming to apply these approaches and their potential improvements to product development (e.g. Millard, 2001, McManus, 2005). For example, one claimed improvement of value-oriented approaches is a 50-90% reduction in wasted time (McManus and Millard, 2002). Nevertheless, existing literature, proposing tools, techniques, and identifying lean principles, lacks a systematic representation and does not practically address the difference between manufacturing and product development (e.g. Browning, 2000). Furthermore, definitions of value in product development indicate aspects of 'goodness', such as flawless product, minimum cost and shortest schedule (e.g. (Slack, 1999), (Beauregard *et al.*, 2008) and (Womack and Jones, 1996)), but often do not give explicit direction regarding how value can be added or measured in this specific context.

In order to broaden the understanding of value in lean product development, its dimensions, i.e. *definition*, *creation* and *delivery,* need to be further explored in a model that:

- relates the dimensions of value;
- helps to understand value dimensions in various contexts;
- considers the impact of different phases in the product life cycle.

These requirements for the model were identified from an analysis of literature of lean in product development. Each requirement will be discussed in greater detail in the following sections.

2.2 Models of Value in Lean Product Development

Research into value in lean product development is relatively young; few models to deepen the understanding of value have been proposed (see e.g. Chase (2001) and Browning (2003) for examples). Most of these models have in common an emphasis on one aspect of value, mainly its creation, but they do not provide sufficient examples to guide application in practice. For instance, Chase (2000) decomposes value into four key layers, which are: *perspective*, *entity*, *attribute*, and *metric*. The first layer is the value *perspective*, identifying to whom value is delivered, such as customer (end user), organization, and stakeholder (e.g. employees and shareholders). The second layer is the value *entity* which produces value for the system drivers (Browning *et al.*, 2002), such as activity, information and resources. On the third layer, Chase odopted Slack's proposed *attribute* for specifying value (Slack, 1999). Main attributes of value include: quality, time, cost and risk. On the fourth layer (*metric*), these attributes can be further analysed in terms such as meters and seconds. These metrics are suggested as performance measures for determining value level in product development processes (Slack, 1999).

A literature review (Siyam 2014) revealed that there is limited common ground to understand, manage and assess value. Therefore, a more holistic model to synthesise the results of the review, clarify the value delivery mechanism and to provide examples that guide application, is necessary. To meet this end, Siyam (2014) introduced an organising model, which will be adopted in this study and discussed in different contexts and phases in the product life cycle. The model aims to (1) synthesise the current understanding, management and assessment of value in lean product development, and (2) provide examples to guide application in practice.

2.3 Value Cycle Model

The Value Cycle Model defines three dimensions facilitating the understanding and improvement of value in lean product development. These dimensions are:

- definition,
- creation, and
- delivery.

The definition dimension determines 'what is considered valuable to whom?'. This includes the identification of stakeholders, such as user, internal customer and shareholders, and their perception of value. The creation dimension explores entities that add value and the mechanisms in which value is added. The main question tackled by this dimension is 'what creates value?'. Finally, the delivery dimension is concerned with measuring entities that carry value. The question here is 'how can value be measured?'.

Figure 2.1 provides an overview of the model. The model can be viewed as a cycle, in which value is added based on requirements set by the stakeholders in the 'definition' dimensions and are continuously assessed to ensure they satisfy the stakeholders. The cycle closes when value is delivered to the 'definition' dimension to confirm a successful value system. The three dimensions are related because a consensus on value understanding, its management and assessment must be reached. Each of the three dimensions is discussed in the next subsections.

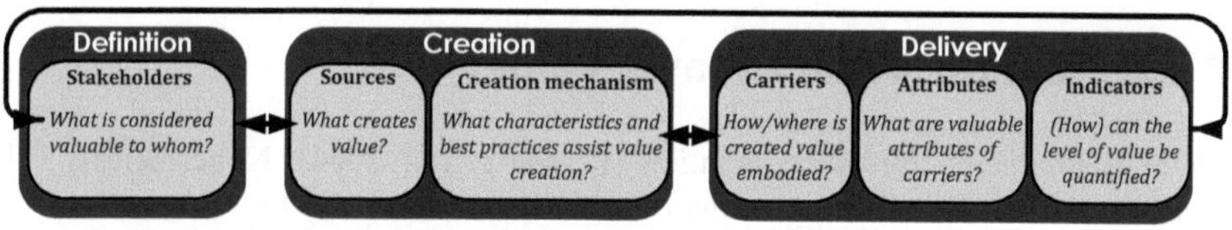

Figure 2.1. Overview of the Value Cycle Model (adopted from Siyam, 2014)

2.3.1 Definition: What is Considered Valuable to Whom?

The Oxford Dictionary defines 'value added' as "the amount by which the value of an article is increased at each stage of its production by the firm or firms producing it, exclusive of the cost of material and bought-in parts and services". The UK Lean Aerospace Initiative team defined value in lean product development as a "capability delivered at the right time, for the right price, as defined by the end user" (Millard, 2001). Browning (2003) suggested that it is more than the value of individual activities, and that it is driven by: 1) the necessary activities and 2) the way they interact to use and produce the right product at the right time.

The definition of value is dependent on the recipent of value, i.e. stakeholders. Stakeholders include users (recipients) of the final deliverable who become involved later in the product life cycle (McManus, 2005). Analysing and understanding the recipients and their needs is critical because they determine the ultimate success of the product in the market. Moreover, the definition of "value" can differ from one perspective to another, which causes challenges in determining priorities and making trade-offs. For instance, with respect to customer value, product performance may be the key determinant of value; however, for some organizations minimising the production lead time may be of higher importance.

In general, understanding value in product development requires consideration of four main perspectives (e.g. (Slack, 1999)). The First is the customer, which is similar to the 'customer' definition in manufacturing, and is composed into two types. The first type is the team member in the next phase or activity. Understanding interactions between team members and controlling them can facilitate value maximization at the process level. The second type of customer is the ultimate user (*recipient*) who receives value through the product features. In that, Browning *et al.* (2002) proposes that the user value is driven by process and product attributes (e.g. cost, delivery, and performance) and depends on customer preference and alternatives. Shareholders form the second category of recipients of value through economic gain or market share (Higgens, 1998). Thirdly, value is delivered to an organisation. Browning suggests that an enterprise receives value by developing and sustaining competitive capabilities and by learning and adapting to become more capable. Finally, employees receive value from compensation, interesting work, and career advancement (Beauregard *et al.*, 2008).

2.3.2 Creation: What Creates Value?

There is no recipe for value creation (Pessôa *et al.*, 2004). Nevertheless, value creation can be described in terms of activity productivity, the information they create, the smooth flow of combined activities, or combinations of those (Chase, 2000). In the creation role, two main issues should be analysed: entities that add value and mechanisms to add value.

Sources: Entities that add value

Literature suggests that three main types of entity add value. The first entity that adds value is *people*, which includes knowledge assets and management. Knowledge assets create value due to their skills and experience which enable actions that generate information contributing to design realisation. Management adds value due to its capability to implement strategies to support improvement and to effectively utilise resources (Penrose, 1959).

The second source of value is the process. Processes are aggregations of development work that can add value when their deliverables meet requirements.

Finally, value is added by methods and technology, which refers to tools, technology and techniques adopted across the product lifecycle, such as product lifecycle management systems. Methods and technology add value as they enable activities to generate information and facilitate the production of information, which may not be possible otherwise. They add value indirectly to the product under development, because they reduce the development time and costs.

Mechanism to add value

Lean in product development suggests different practices and approaches to most effectively add value. Value approaches discussed in the literature include, for example, "having a core team" (for people) and "establishing takt time" (for a process). Examples of these practices are summarised in Table 2.1.

Table 2.1. Examples of value practices

Entity	*Lean practices (e.g.)*	*Reference (e.g.)*
People	*Assemble a core team*	Oppenheim (2004)
	Ensure top management involvement	Eskerod (2009)
Process	*Define value, e.g. using a value break-down structure*	Pessoa (2007)
	Sequencing activities to maximize value	Browning (2002)
Methods & Technology	*Promote awareness of cost*	Browning (2002)
	Manage inventory effectively	Reinertsen (2009)

2.3.3 Delivery: How Can Value be Measured?

The value-adding entities and the value they create have several attributes that can be quantified and used later for analysis, control, or improvement (see Table 2.2.). Browning suggested that performance, risk, schedule, and cost of developing design are the main attributes that characterize value. Slack decomposed basic attributes such as cost, performance, and timeliness (Slack, 1999). In contrast, McManus suggest that product development primary consists of information flow and chose form, fit, function, and timeliness. Oppenheim (2004) proposes that value in terms of quality include: mission assurance product integrity, life cycle performance, first time quality, safety, redundancy, functionality, robustness, durability, flexibility, maintainability, sustainability, support, and other customer requirement characteristics. Table 2.2 gives some examples of indicators used to assess value in product development.

Table 2.2. Examples of value metrics

Entity	*Attribute*	*Indicators (e.g.)*
Information	*Certainty*	1/ Range of possible values for a design parameter
Knowledge asset	*Experience*	Percentage of engineering errors or unplanned iterations
		Processing time
Process	*Time*	Actual time/planned time
Delivered product	*Customer satisfaction*	Market share
		Sales percentage

2.4 Value Cycle Model in Context

"Value has a connotation which is singularly personal; it depends on the evaluator, his viewpoint, and the prevailing circumstances." (Asimov, 1962)

The understanding of the value cycle is dependent on the individual perspective of a stakeholder. People that can have an interest in an project are for example involved designers, project manager, corporate management and shareholders but also customers, competitors, or even the society as a whole and government. The different stakeholder groups and factors that have an impact on a project are the context of product development. Factors can have a direct or an indirect influence on product development as they are strongly interdependent. Different schemes for representing the context of product development exist (Gericke *et al.*, 2013). Here, context means influencing factors, i.e. "'people or things having power,' with power as 'the ability to affect outcomes'" (Hales and Gooch 2004, p. 29, referring to Lawrence and Lee, 1984) The term context factor is used synonymously with influencing factor, i.e. a factor having an influence on a project.

Hales and Gooch (2004) provide a list of context factors and propose five levels of resolution to structure them: *macroeconomic, microeconomic, corporate, project, personnel*. An extension of this list is presented by Gericke *et al.* (2013).

Linking the value cycle model with Hales and Gooch's context model (see Figure 2.2) allows an illustration of:

- the specific understanding of value on each context level; and
- the hierarchical relationship between the different context levels, i.e. the top-down relationship for defining goals, thus defining the understanding of value and the bottom-up relationship for managing value creation and assessment of the deliverables.

An illustrative example is the following. Based on a market analysis, the management of a company initiates a new product development project. Stakeholders in this project include the management itself, shareholders of the company, the project team, and customers. The management will define goals for the product development project, which will guide downstream activities including product development, manufacturing, sales, etc.

During the product development project, work packages will be assigned to designers. For each activity and each role (stakeholder category) specific lean practices are available. In addition, specific indicators for attributes for assessing whether the carriers satisfy previously defined goals for value creation do exist.

While goal definition is mainly done top-down, the creation of value is done bottom-up and assessed by a superior level in the context model. That means the team and the project leader will assess the activity of an individual designer. The management will assess the deliverable of the team and the shareholders will assess the contribution to the company growth. Customer and competition will assess the developed product.

For effectively modeling the *definition*, *creation* and *assessment* of value in the context of lean product development, the value cycle has to be analysed on each level of resolution of the context considering the top-down and bottom-up relationships between the different levels, as shown in Figure 2.2.

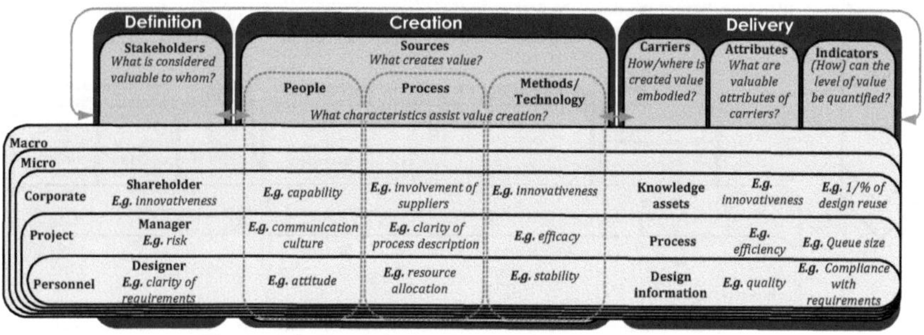

Figure 2.2. Extended Value Cycle Model in Context

2.5 Value Cycle Model with Respect to the Product Life Cycle

When widening the scope of the analysis of value with respect to the product life cycle, the relevance of considering individual perspectives and considering context dependency become even more important than when considering the product development phase alone. The transformation of the value cycle along the product life cycle is not sufficiently articulated in the literature, even though it is one of the main influencing factors determining the understanding, creation and assessment of value. Neglecting this influence easily results in the perception of inconsistency across value approaches that exist for specific life cycle phases.

The Value Cycle Model can be used to explain the perspectives of different stakeholders along the entire product life cycle (*understanding*). The logic will remain the same irrespective of the particular product life cycle phase (see Figure 2.3). However, sources and lean practices that support value creation (*manage-*

ment) and metrics that are appropriate to assess value delivery (*assessment*) will be different for each life cycle phase and may differ depending on the chosen perspective (context level).

The application of the Value Cycle Model in practice requires analysing the time-dependency of the value cycle, i.e. specific entities have to be identified, and applicable lean practices and metrics need to be selected.

Depending on a particular product life cycle phase activities differ and different stakeholders typically become more prominent. In consequence different stakeholders will have a higher impact on value creation (Gericke et al. 2012), as shown in Figure 2.3.

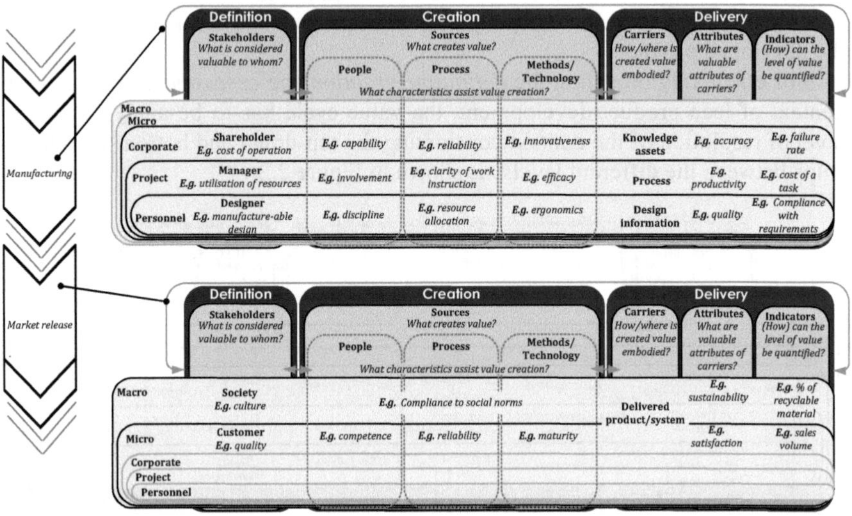

Figure 2.3. Evolvement of the Value Cycle Model along the Product Life Cycle

2.6 Discussion

The Value Cycle Model is intended to support the understanding, thus improvement of value in lean product development. This is by articulating definitions of value in lean product development and clarifying the value creation mechanism, which as a result can assist in improving the value system. For example, to optimise value creation using this model, one first starts with identifying the stakeholders and what their preception of value is. Afterwards, one can define sources of value and what should be done to achieve the requirements of stakeholders. This is then linked with a set of measurements to continuously assess the value level and to ensure stakeholders are satisfied. In this paper, the Value Cycle Model (see Figure 2.2 and 2.3) was extended by exploring it in:

- different levels of resolution of the context (illustrated in Figure 2.2), and
- a life-cycle view (illustrated in Figure 2.3).

The introduced dimensions highlight the importance of understanding value with respect to a particular context and product life cycle phase, which should be analysed and improved with regards to value creation. The model explicitly considers:

- the end-to end relation of the value process
- that characterisitics of the value cycle are specific for each life cycle phase, i.e. value creation in product development is different from manufacturing
- that multiple dimensions of value exist for different stakeholders depending on their specific view, i.e.:
  - the top-down and bottom-up relationship in understanding, creating and assessing value,
  - the assessment of value can e.g. focus on the product or system that is delivered, its contribution to the economic wealth of a company or its life-cycle value.

Some limitations still remain in this model. First, the model shows the aspects of value creation to be equally important with a linear relationship. However, the interaction between these elements and their relative importance is a critical issue that cannot be overlooked when analysing value.

Second, while time is already considered in the extended Value Cycle Model as an important influencing factor shaping the whole Value Cycle System, an important dimension of value creation is currently not completely considered, namely dynamics. Taking this issue into account it is clear that the time dependency is not as simple as depicted in the extended Value Cycle Model. The dynamics results for example from changing stakeholders involved along the product development process, but also due to changes of the particular importance of their influence on the process as well as changing criteria for assessing value creation in different phases of the product development process. Currently, only the dynamic change of the value cycle along the product life cycle i.e. the inter-life-cycle phase dynamic is considered in the model. The intra-lifecycle phase dynamics, i.e. the dynamic changes of the value cycle within single life cycle phases is not illustrated.

Considering the dynamic will not necessarily increase the complexity of the Value Cycle Model, as it allows a time-dependent prioritisation, i.e. only a subset of stakeholders, sources, carriers, and metrics, are relevant in a particular context and life cycle phase.

2.7 Conclusions

Value is understood, managed and assessed differently in product development and manufacturing. A common definition of processes, creation mechanisms, and measures is important to enable greater focus on work that creates most value. Therefore, this paper aimed to discuss the Value Cycle Model relative to different contexts and product life cycle phases in lean product development. The extended Value Cycle Model integrates different views on and notions of value in product development. It provides an overview about the dependencies between understanding, managing and assessing value creation in product development. In its current version the model establishes a logical and causal connection between relevant entities. It is intended to support focusing on particular entities involved in PD value creation while emphasising that they are embedded in a network, thus highlighting dependencies, which have to be considered when addressing value in product development.

2.8 References

Asimov, M. (1962) Introduction to design, Prentice-Hall

Beauregard, Y., Thomson V., Bhuiyan, N. (2008) Lean engineering logistics: load leveling of design jobs with capacity considerations. *Canadian Aeronautics and Space Journal*, Vol. 54, No. 2, pp. 19-20, DOI 10.5589/q08-006

Browning, T. R., Deyst, J. J., Eppinger, S. D., Whitney D. (2002) Adding Value in Product Development By Creating Information and Reducing Risk, *IEEE Transactions on Engineering Management*, Vol. 49, Issue 4, pp. 443-458, DOI 10.1109/TEM.2002.806710.

Chase, J. P. (2001) Value Creation in the Product Development Process, Massachusetts Institute of Technology, Masters of Science in Aeronautics and Astronautics

Eskerod, P. (2009) Project Management Models as Value Creators, *Project Management Journal*, Vol. 40, No. 1, pp. 4–18, DOI 10.1002/pmj.20098

Gericke, K., Meißner, M., Paetzold, K. (2013) Understanding the Context of Product Development; *in Proceedings of 19th International Conference on Engineering Design, Seoul, Korea*

Gericke, K., Blessing, L. (2012) An Analysis of Design Process Models Across Disciplines, *Proceedings of the 12th International Design Conference*

Hales, C., Gooch, S. (2004). Managing Engineering Design. 2nd ed. London: Springer-Verlag. Hollins, B., Pugh, S. (1990). Successful product design: What to do and when. London: Butterworths

Higgins, R. C. (1998) Analysis for financial management, Irwin/McGraw Hill

Lawrence, P. A., Lee R.A. (1984) Insight into management, Oxford University Press, Oxford

McManus, H. L. (2005) Product Development value stream mapping (PRODUCT DEVELOPMENT VSM) manual, Lean Aerospace Initiative, Release 1.0

McManus, H. L., Millard R. L. (2002) Value stream analysis and mapping for Product Development, *Proceedings of the international council of the aeronautical sciences, 23rd ICAS Congress, Toronto, Canada, 8-13, September*

Millard, R. L. (2001) Value stream analysis and mapping for Product Development, Masters of Science in Aeronautics and Astronautics, Massachusetts Institute of Technology, June

Oppenheim, B. W, (2004) "Lean Product Development flow," *Systems Engineering, Vol.7, Issue 4, DOI 10.1002/sys.20014*

Penrose, E. T. (1959) The theory of the growth of the firm, John Wiley, New York

Pessôa, M. V. P., Loureiro, G., Alves, J. M. (2007) An approach to lean Product Development planning, In: *Complex Systems Concurrent Engineering*, Part 4, pp. 229-237, DOI 10.1007/978-1-84628-976-7_26

Reinertsen, D. (1999) Lean thinking isn't so simple, *Electronic Design*, Vol. 47 Issue 10, p48H, May

Slack, R. A. (1999) The lean value principle in military aerospace product development. The Lean Aerospace Initiative, Report Series, PR99-01-16

Shannon, C. E. (1948) A mathematical theory of communication, *The Bell System Technical*

Journal, Vol. 27, pp. 379–423, 623–656
Siyam, G. (2014) Value in lean product development, Thesis, University of Cambridge, Engineering Department
Siyam, G., Wynn D. C., Clarkson P. J. (2011) Re-conceptualising value in the engineering design process: The Value Cycle Map, in 18th International Conference on Engineering Design (ICED'11), Copenhagen, Denmark
Womack J. P., Jones D. T. (1996) Lean thinking, Simon Schuster, New York

Chapter 3

Case Study on Requirements Management in Multidisciplinary Product Development

A. Albers, E. Wintergerst, T. Pinner and J. Breitschuh

3.1 Motivation and Research Background

This paper reports about a case study on documentation and communication workflows that support requirements management in a company that is a global supplier for automotive and industrial engineering projects. The scientific motivation behind this case study was to explore how individual project managers cope with the *generation of a validated set of requirements*. Second, the case study was focussed on processes and tools that support a flexible *documentation and communication of requirements*. Requirements in this context were defined as qualitative and quantitative parameters (VDI 2221) that refer to the expected characteristics and properties of the products (Weber, 2013).

The intention for this publication is drawn from the demand of design research for empirical findings from design practice. Instead of discussing recent scientific approaches, we would like to focus this paper on sharing our observations and provide insights into industrial engineering practice for the design research community. This is intended to provide an empirical data basis on which theoretical studies about requirements management can be built on.

The following paragraphs thus outline the methodology that was used to record and to analyse the requirements management processes. Results will be presented and discussed in comparison to the individual characteristics of the case study projects.

The case study was conducted in a company that is facing a rising pressure to facilitate requirements management in multidisciplinary development projects because of a changing market situation. Today, most of the company's products can be considered as single mechanical machine elements or mechanical subsystems that are being integrated into more complex machines which are developed by its customers. Due to an increasing demand for system suppliers, the company is expected to develop more and more complex mechatronic assemblies

M. Schabacker et al. (eds.), *Modelling and Management of Engineering Processes*, DOI 10.1007/978-3-662-44009-4_3

in future years. As a result, its products will be characterized by an increasing number of integrated mechanical, electronic and software-based components.

This situation imposes new challenges for the company. Its standard processes, methods and tools for requirements management were defined for developing mechanical machine elements or subsystems, but were suspected to not efficiently support more complex multidisciplinary development projects. However, some business units have already made experiences in managing requirements for more complex mechatronic assemblies. They were expected to be able to report about challenges, needs and best-practices in a retrospective analysis of their development projects. Thus, the company decided to initiate this case study and to provide a group of researchers access to four representative product development projects in four of its business units.

3.2 Setting of the Case Study and Research Methodology

The aim of the case study was to explore how the individual development teams cope with the generation, actualization, communication and implementation of requirements into the workflow of their departments. For this purpose, the four product development processes were recorded retrospectively from the beginning of development up to the product's start of production.

3.2.1 Characteristics of the Reference Projects

Each of these four product development projects belongs to a different business unit of the company: Tooling Machines, Wind Power Systems, Laboratory Equipment and Power Train Systems. Each business unit comprises multiple departments e.g. for design, calculation/simulation, production or prototyping and partly own departments for project management as well.

Project I – Tooling Machines: The business unit Tooling Machines develops and supplies machine tools for prototypes as well as for mass production within the case study company. The two participating departments in this project were both working on mechanical design problems from different perspectives: production system development (contractor) and product development (client). The final product of the tooling machine development process is a single item (one tooling machine) but no mass product (like the products that are manufactured with it). Every participant in this project belongs to the case study company; there are no direct interfaces to external customers.

Project II – Wind Power Systems: The output of this project is a small subsystem that is produced as a small batch series and integrated into a wind turbine. The customer company is the manufacturer of this wind turbine. The four internal specialist departments that were mainly contributing to the requirements specification (mechanical engineering, condition monitoring, quality management and computation/simulation) were working within the same business unit, each taking an individual perspective in the project. A central project management team was organizing the documentation and communication workflows within the business unit as well as with the external customer company.

Project III – Medical Devices: This project was established by the company's own initiative, but with an external, potential customer in the background. The first concepts were thus developed as a prototype according to his requirements. Later on, further adaptions were carried out on the product in order to supply multiple potential customers with prototypes. Finally, a product development process was performed that ended with the start of mass production. The product itself is a small sub-system of medical laboratory equipment. It includes mechanical as well as electrical components. The five mainly participating specialist departments (two different mechanical engineering and one electrical engineering departments, testing and customer services) belong to two business units, sharing their knowledge in mechanical and electrical product development in order to include new product functions into their portfolio.

Product IV – Power Train Systems: The product that was developed in this case study project was ordered from an automotive company (OEM) for mass production. It can be considered as a complex mechatronic system, including mechanics, electronics and software components. Since the customer company was at the beginning of its own product development process, intensive communication was initially required in order to reconcile requirements between the external customer and internal specialist departments. Six of these specialist departments were participating in this case study, representing the development, testing and sales departments as well as the central project management.

The individual complexity of requirements management depends on the above described characteristics of the case study projects (cf. Table 3.1). In this context, complexity is considered to be defined by the number of participating disciplines and specialist departments, by the ratio between internal and external sources of requirements as well as the number of multidisciplinary interfaces between sub-systems of the product. Considering these factors in comparison between the four case study projects, the business units were facing different challenges for requirements documentation and communication which will be discussed in detail in section 3.3.

Table 3.1. Characteristics of the case study projects

Project No.	**I**	**II**	**III**	**IV**
No. of departments involved in product development	2	4	5	6
Mechanics development involved	✓	✓	✓	✓
Electronics development involved	-	✓	✓	✓
Software development involved	-	-	-	✓
Main source of requirements (i = intern, e = extern)	i	e	i,e	i,e
Complexity of the requirements management	low	moderate		high

The four projects' common ground is their affiliation to the main company and, thus, having some paramount process elements in common. Otherwise, the projects differ significantly concerning their system's complexity, the relevant engineering domains (mechanical, software and electronics), the customer relationship, the amount of stakeholders and developers and the final output lot size. Hence, the development projects are considered to represent a wide range of the company's manifold product portfolio. However, interpretation of this case study's results considers the unique character of each business unit and development process as well as the overall low amount of samples. Results of this case study are thus considered as best-practice examples rather than universally valid conclusions. Nevertheless, they may be suitable to improve the requirements management when transferred to a neighbouring business unit and applied under similar conditions.

3.2.2 Research Methodology

The case study was structured in two work packages: (1) preparation and (2) investigation. They focus requirements management activities from two different perspectives in order to better understand characteristics of the case study projects that may influence the requirements documentation and communication strategies.

For preparation purposes, the products as well as the corresponding development processes were examined.

First, functional and structural decomposition and analysis of the four reference *products* were performed. The aim was to identify key components of the products that were suspected to be of special importance for the product's main functions – and thus of special interest for requirements management in the corresponding development process. For this purpose, Design Structure Matrices (DSM, cf. Lindemann, 2009) were used to identify which components directly or indirectly contribute to the product's functions. According to the active sum of each line in a DSM, each component was rated and key components were selected. The analysis was based on assembly drawings of the products as well as on interviews that were focused on the wirk-structure of the products.

Second, the corresponding development *processes* were recorded and modelled. The aim was to survey wheter the key components that were identified in the DSMs were especially considered in risk assessment, project planning and requirements management. The data was collected in one-day workshops with the project managers and representatives of the specialist departments based on the Metaplan technique and semi-structured interviews. The Metaplan model depicts traceable objects (like documents, email, meeting minutes) over the entire development process. For analysis and comparison purposes, all four development process models were consolidated in a large swim lane representation. As a result, it could not be observed that any of identified key components were especially considered in risk assessment, project planning or requirements management.

For investigation purposes, a second series of workshops was performed in order to focus on requirements management activities. The background-knowledge about the products and the corresponding development processes was used to establish a semi-structured questionnaire which addresses individual documentation and communication workflows in the case study projects. One individual workshop was performed for each case study project. Each was moderated by two researchers that conducted semi-structured interviews based the above mentioned questionaires. Participants in each of these four workshops were the individual project manager as well as one to five development engineers from the specialist departments that were involved in this project. Conclusions have been derived by comparing the interview protocolls of the different projects. Results of this investigation are presented in section 3.3.

3.3 Case Study Results

This section is dedicated to selected results of the case study. It gives an overview on processes and tools that were used to specify and to communicate requirements in cooperation with external customers and internal specialist departments. In this context, one focus is set on the initial process of requirements specification and verification. Another focus is set on internal documentation and communication strategies that support information about updated specifications.

3.3.1 Requirements Acquisition

A common situation for all projects participating in this case study was an initially incomplete definition of requirements. Therefore, the project organization was arranged around a central project manager or management team whose first task was to exchange and to consolidate technical specifications with the customer company and internal specialist departments. Depending on the degree of *maturity* and *rigidity* of the initial set of requirements, the first engineering activities could be focused either more on requirements acquisition and verification or on conceptualization and validation activities. According to (Albers *et al.*, 2011), "the degree of maturity describes the completeness regarding the understanding and realization", while "the degree of rigidity indicates the willingness to hold on" to a certain requirement.

In project I, the initial requirements could be defined with high degrees of maturity and rigidity. Since the customer was located within the same company (but in a different business unit), they were sharing the same documentation and communication standards: production drawings with explanatory notes. The initial specification sheets could thus easily be exchanged and implemented into the receiver's development workflows. It was the responsibility of the delivering business unit to check the validity of requirements that were specified by the production drawings against their standard design guidelines.

In project II, a comprehensive specification sheet was initially delivered by the external customer company. The first requirements management task was thus to examine the completeness, validity and feasibility. For this purpose, an initial product concept model was established. Based on calculation results, experienced data and an internal design guideline, a requirements deviation report was derived together with internal specialist departments. After being approved by the customer, the deviation report and the initial specification sheet were merged into a central requirements specification that was used as a basis for product development.

The initiative for project III was taken from the case study business unit based on observations from the customer support department. First requirements specifications were derived from the personal contact with one potential customer and recorded in a central requirements list. They were thus rated with a medium degree of maturity and a low degree of rigidity. For further clarification purposes, information was informally exchanged via telephone, e-mail and personal meetings until a fixed set of requirements was accomplished as a basis for an advanced engineering project (AEP). New customers could be acquired based on the AEP results. They submitted a request for a proposal which included detailed specifications. The completeness, validity and feasibility of these requirements could be examined against the experienced data from the AEP and informally approved deviation proposals with the customer.

In project IV, the initial requirements specifications were provided by the customer company and exchanged as lists via e-mail. Since the customer company was at the beginning of product development itself, these specifications were rated with a low degree of maturity and a low degree of rigidity. They were focused only on the main product properties, included just a rough summary of boundary conditions and were considered to be subject of frequent change requests. In the early stages of this project, almost all engineering activities were therefore focused on requirements clarification and verification. Because of the low degree of maturity and rigidity, the initial specification sheets had to be developed iteratively and simultaneously to the first product concepts. They finally served as a basis for a technical proposal and a complementary quotation to the customer company. Due to the multidisciplinary character of the project, multiple fragmented documents were used to record and to communicate initially defined requirements (cf. Table 3.2). After the project was officially established, all requirements that were rated with a high degree of maturity and rigidity were transferred into a central requirements management software tool (cf. section 3.3.2, Table 3.3).

Table 3.2. Characteristics of initial requirements acquisition workflows

Project No.	I	II	III	IV
Initial degree of requirements maturity	↑	↑	→	↓
Initial degree of requirements rigidity	↑	↑	↓	↓
Communication of initial vague requirements	f	f	i	f,i
Documentation of initial vague requirements	c	c	c	fr

↑ = high, ↓ = low, → = medium, f = formal based on documents, i = informal based on telephone/e-mail, c = central in a project database, fr = multiple fragmented sources

During the case study interviews, it could be observed that the degree of maturity of the initial set of requirements did not influence decisions of the project managers which type of documentation and communication means to select. For early stages of product development, a document-based requirements communication strategy was preferred in all case study projects.

3.3.2 Continuous Requirements Documentation and Communication Strategies

Compared to development processes for mechanical machine elements, the increasing variety of interfaces between mechanical, electronic and software-based components affords a significant intensification of communication between the internal departments that are involved in the corresponding development processes. It was thus expected that adequate types of documentation were selected in the case study projects that have a positive influence on the requirements communication workflows. In order to test this hypothesis, semi-structured interviews were performed with the project managers. In the following paragraphs, the observations will be outlined in detail.

In project I (Tooling Machines), production drawings were the most important documents concerning the recording of product characteristics. A supplementary set of documents was used to record product properties specifications and approved deviations from initial requirements specifications. Change requests were informally handled via telephone or e-mail communication. All requirements specification documents were stored in a central project database in the company's IT-network. The expertise of the development engineers complemented the information in these documents. This refers not as much to the extent of completeness, but more to the relationship between product properties and characteristics. Functions were not explicitly recorded in either of these documents.

In comparison to this document-based, decentralized requirements management strategy, in project IV (Power Train Systems) all participating departments could access product requirements in a central requirements management software tool. However, development engineers as well as the project management were deriving text-documents from their central set of requirements for internal and external communication purposes, e.g. to capture notes and to focus meeting discussions on a selected set of specifications. The requirements management tool was thus more used as a documentation tool for already fixed specifications rather than being implemented in the communication workflow. This document-based strategy was also applied to exchange information about requirements with the customer, e.g. to collect new and to sharpen initially vague specifications during multiple iterations. Change requests from the customer side and internal specialist departments were recorded and exchanged in single standard forms (table and text files) by the central requirements management team.

Project II and III were following a document-based and personally centralized documentation strategy. Requirements were initially specified in one text-document which complemented the relevant production drawings (cf. Table 3.2). It could be accessed via SAP or the central project database in the company network. For consistency and safety reasons, this central text-document could only be modified by the project managers. It was subject to version control procedures and needed to be archived on a regular basis. Change requests were informally reported to the project managers. In the course of product development, single specialist departments were deriving further specification sheets for internal communication purposes. As a result, the project managers had to cope with an increasing fragmentation of requirements documentation.

In project II (Wind Power Systems), four different department-specific documents were derived from the central requirements specification list for internal communication purposes: an *Engineering Specification*, a *Deviation Report* based on the customer specifications, a *Design Verification Plan* as well as a *Mounting and Operation Specification.* According to the project management, this redundancy became necessary throughout the project, because the large number of requirements could not be recorded in a single text-document structure that allowed a fast summary of relevant information. As a consequence, the project manager was responsible for keeping all documents constantly updated and consistent throughout the project.

In project III (Medical Devices), one central text-document was used to record requirements specification for internal reference and communication purposes. This document was complementing a set of production drawings that included detailed specifications about product characteristics. To exchange and to contract specifications with the external customer company, copies with less detailed information were derived. For certain internal development activities such as design verification tests, the development department was ordering services from specialist departments. Results were reported in standard forms, including recommendations for updated specifications and explanatory statements. In the course of the project, this communication strategy led to an increasing fragmentation of documents that include explanations about requirement change procedures.

Table 3.3. Characteristics of requirements documentation and communication strategies

Project No.	I	II	III	IV
Communication of fixed requirements	f	f	f	f
Documentation of fixed requirements	c	fr	c	c
Communication of requirements change requests	i	i	i	f
Documentation of requirements change procedures	fr	c	fr	c

f = formal based on documents, i = informal based on telephone / e-mail, c = central in a project database, fr = multiple fragmented sources

Analysing the type of requirements based on the CPM-approach (Weber, 2013), it could be observed that the application of a decentralized documentation strategy (e.g. in project I and III) leads to the use of different types of documents for either product properties or characteristics. Product characteristics were preferably documented in CAD-models and complementary production drawings, while product properties were recorded in supplementary documents such as specification lists. In contrast, a centralized documentation strategy (in project II and IV) leads to a joint recording of product properties and characteristics according to the topics that they address, e.g. mechanical design, validation processes, maintenance and operation modes etc.

The ability to access a commonly shared IT-network was observed to have a major influence on the documentation strategy. While company-internal departments could exchange documents via a central project database, the dialogue between customers and the project management was bound to e-mails and document attachments such as specification lists, meeting notes and telephone protocols.

Although the initial process of requirements specification had been carried out carefully, all participating project managers reported in this case study that requirements had to be constantly updated, added and prioritized throughout the entire development process. The initial set of specifications and all following requirements documentations were thus undergoing frequent revisions in order to reassess risk analysis and project management calculations (i.e. project costs or resource planning). In order to prevent fragmentation, discrepancy and to be protected from any unnoticed changes or manipulations in requirements documentation, all project managers preferred a single-source strategy: in project II and III, only the project manager was allowed to define new requirements or changes to existing requirements within the shared documentation. In project I, requirements were recorded in production drawings which could only be edited by designated development engineers from the product development department. In project IV, a comprehensive documentation of all fixed requirements was accessible via the central requirements management system.

The decisions of the project managers, which type of documentation to select for recording and exchanging requirements, were guided

- on the one hand by the aim to achieve a suitable degree of formalism, structure, transparency and thus security in requirements management,
- on the other hand by the aim to optimize documentation and communication workflows.

However, the subjective degree of efficiency in managing requirements was rated different in the case study projects.

In project I, development engineers and the project management did not indicate any need for a more advanced requirements management system. In contrary, they suspected that it could disturb the efficiency of well-established communication workflows within and between the product development and the production system development departments.

The project managers of project II and III reported that the multiple-documents strategy to record and exchange requirements could basically satisfy their needs for communication with customers and the internal specialist departments. For future projects, a growing need for multidisciplinary cooperations and thus more complex interrelations between requirements were suspected. The project managers thus indicated that a more advanced requirements management system should facilitate evaluating consequences of changes or automatically reporting changes to the departments that are affected.

The efficiency of a document-based requirements communication strategy was confirmed by the management of project IV for clarification purposes in early stages of the project. However, once a set of requirements with a high degree of maturity and rigidity could be fixed, a central requirements management tool was preferred over a document-based communication strategy, e.g. to assist in consistency analysis and change management. Both in project III and IV a central requirements management tool was rated to be more effective to support communication workflows in case many departments were involved and a common agreement about a basic set of requirements could have been achieved.

3.4 Discussion and Outlook

It was observed in this case study that requirements management can be understood as a backbone for the product development process. The initial requirement acquisition is setting a fundamental basis for managing a development process, e.g. referring to decisions how to efficiently provide access to requirements, how to arrange communication workflows between specialist departments and which activities to perform in order to create first product concepts. The following paragraphs will discuss these three issues considering the individual character of the four case study projects.

The type of documents and tools to record and to exchange requirements were selected and adapted according to the number of participating departments as well as their preferred communication workflow processes. In project I, the participating departments were both working on mechanical design problems from different perspectives (product and production system). Thus, they preferred to use similar tools and types of documents for the exchange of requirements. The communication workflow was centred on production drawings that could be exchanged without information restrictions because both parties belonged to the same company. In project II and III, a strong focus was set on mechanical design problems as well, but they were accompanied with multidisciplinary problems that could only be solved together with the external customer company and a low number of internal specialist departments. As a result, the project managers preferred a document-based and centralized documentation and communication strategy: one text-document as well as a small set of production drawings was used as a basis to derive sub-documents representing the specific, limited scope of each communication process. Project IV was characterized by the highest degree of product complexity regarding the number of interfaces between sub-systems and the number of participating disciplines and specialist departments. While requirements acquisition was initially performed with multiple fragmented documents, it was necessary to introduce a central requirements management tool for the further development process to cope with the high complexity of requirements management.

In the case study projects, the rising complexity of requirements management tends to result in a

- comparatively lower degree of maturity and rigidity of the initial set of requirements,
- more intense informal communication for requirements acquisition
- more fragmented documentation of requirements if a central requirements management tool cannot be used
- more formal communication of change requests and a more formal documentation of change procedures.

To support communication between the specialist departments, all project managers preferred a single-source strategy to provide access to requirements documentation. Especially for early stages of product development, a document-based requirements communication strategy was preferred over a central requirements management tool. The same applies to the requirements clarification dialogue between customers and the project management, even in later stages of product development. However, with a growing network of interrelated requirements and growing degrees of maturity and rigidity, a central requirements management tool was estimated more effective than a document based approach.

Since all case study projects were characterized by an initially incomplete definition of requirements, clarification and verification were the dominant activities in early stages. For projects with an initially high degree of requirements maturity, product development activities could be focused earlier on activities of solution synthesis rather than on analytical clarification and verification activities. This observation underlines the significance to establish a comprehensive and valid initial set of requirements at the beginning of product development.

Future research activities will be focused on analysis of the case study data to examine how different documentation strategies and tools that are used for requirements management may impact on the collaboration workflow.

3.5. References

Albers A., Lohmeyer Q., Ebel B. (2011) Dimensions of objectives in interdisciplinary product development projects, in: proceedings of 18th International Conference on Engineering Design ICED 11, Copenhagen, Denmark

Lindemann U., Maurer M., Braun T. (2009) Structural Complexity Management – an approach for the field of product design. Springer, Heidelberg

VDI-Richtlinie 2221 (1993) Methodik zum Entwickeln und Konstruieren technischer Systeme und Produkte. Beuth, Berlin

Weber C. (2014) Modelling Products and Product Development Based on Characteristics and Properties. In: An Anthology of Theories and Models of Design: Philosophy, Approaches and Empirical Explorations (Blessing, L.T.M.; Chakrabarti, A.). Springer, Heidelberg

Chapter 4

A Model Based Approach to Support Risk Management in Innovation Projects

M. Neumann, M. Sporbeck, T. Sadek and B. Bender

New product development is notably affected by uncertainties that are a consequence of insufficient experience and missing knowledge. If uncertainties are not managed adequately, they will finally lead to risks. We therefore advocate an integrated agile development process, allowing for explicit modelling of uncertainties and reaction strategies as well as the evaluation of the resulting risk caused by the changes to the product in development or the development process. As changes can again lead to undesired change propagation, finally resulting in new uncertainties and in consequence new risks, uncertainty response strategies need to be developed, evaluated and conducted collaboratively. In this publication the Integrated System and Risk Managing Model is presented, enabling users to describe and analyze product and process based uncertainties as well as potential response options within one consistent system. This paper elaborates the underlying structure of the model and concentrates on the modelling process, also explaining the application using examples from a case study.

4.1 Introduction

Product innovation is the result of a renewal process that broadens knowledge or applies available knowledge in a new context (Ericson and Kastensson, 2011). Innovation projects are thus accompanied by the presence of uncertainties that in general are understood as a consequence of insufficient experience and missing knowledge (Ehrlenspiel, 2007). Uncertainties may occur at all stages of product development, potentially influencing the entire product lifecycle (Browning, 1998). It is obvious that uncertainties not handled adequately will seriously affect project success. Managing uncertainties is therefore essential in order to reduce risks in innovation projects.

M. Schabacker et al. (eds.), *Modelling and Management of Engineering Processes*, DOI 10.1007/978-3-662-44009-4_4

Traditional engineering approaches follow a sequential process assuming that the entire system is developed top-down. Verification and validation activities are primarily carried out at the end of the development process. In consequence, uncertainties are addressed behind time when expensive cost and schedule overruns are no longer avertable. Especially the rising complexity of products and the increasing pressure of shortening the feedback loops have stimulated the creation of incremental development approaches. These are based on the idea of subdividing the complex development project into smaller iteration cycles which then deliver fast feedback, for example by providing prototypes with growing level of maturity. Approved incremental models are e.g. the Spiral Model (Boehm, 1988) and the V-Model that both were initially developed for software engineering and later adapted to other industries (VDI, 2004). While these models already cater for a more dynamic proceeding, they still do not address uncertainties as a central problem of new product development explicitly.

In order to address uncertainties and resulting risks in a more thorough manner, specific risk management models were established. These models describe risk management on an operational level as an iterative procedure, usually comprising the stages of risk identification, risk analysis and risk response (Ferreira and Ogliari, 2005). However, risk management commonly coexist beside the superior models of product development.

Due to the particular significance of uncertainties in new product development we ask for an integrated product development and risk management model considering uncertainties explicitly in decision making. Moreover, we propose a highly agile development process for innovation projects enabling immediate reactions to upcoming uncertainties by conducting risk oriented changes to both, the product in development as well as the development process.

The procedure presented in this contribution is supported by an integrated modelling approach based on Multiple-Domain Matrices (Maurer, 2007), enabling the representation and analysis of product and process based uncertainties as well as potential response options. Several response strategies can be evaluated directly with regards to the caused benefit and effort, and thereby made comparable. This paper focuses on the modelling approach and its application in the risk management process. The theoretical background is presented in section two. Section three presents a real-life scenario detailing the challenges of new product development and motivating the proposed method. This is followed by a discussion of related work relevant in the presented context. Section five describes the modelling-approach as well as the underlying procedure, followed by an application to the example given within the case study. Finally, section six concludes the chapter giving an overview about ongoing research related to the presented approach and future work.

4.2 Theoretical Background

In order to understand the presented approach, a clear definition of the key concepts of *uncertainty* and *risk* has to be achieved.

In literature the term *uncertainty* is not universally defined. Several definitions are used, originating from different disciplines of research. *Uncertainty* as defined in decision theory relates to the information base relevant for decision making. A *decision made under uncertainty* can thus be understood as a *decision based on uncertain decision criteria.* These uncertain criteria comprise potential deviations of product or process properties caused by knowledge deficits at the point of decision making (Engelhardt *et al.*, 2011), discrepancies between the information currently available and the information necessary for conducting a task (Verworn, 2005) as well as statistical process results or information not yet collected ("things that are not known, or known only imprecisely" (Hastings and McManus, 2004)). Based on literature review we identified seven classes of uncertainties, namely uncertainties rooted in the market context and use context, in politics, law and society, technology, fabrication, procedure and applied methods as well as the utilized resources. These classes can be further subdivided into endogenous and exogenous types (Weck *et al.*, 2007).

In context of decision making, an additional interpretation referring to the result of the decision process is relevant to our method. A decision based on uncertain decision criteria may be seen as uncertain itself. In order to classify these we propose a categorization aligned to the three partial systems: the target system, the technical system and the execution system which constitute the generic reference frame for the decision process.

The term *risk* is also discussed controversially. Bitz e.g. defines risk simply as a danger of loss (Bitz, 2000). In a similar manner Smith and Merritt describe risk as the hazard of project disruptions triggered by an undesired event or the absence of a desired event (Smith and Merritt, 2002). The extraction of definitions already shows that a clear differentiation between the terms uncertainty and risk often does not exist. This fact can be traced back to the divergent understanding of the term risk, differentiating between a cause based and an effect based interpretation. Following the cause based interpretation risk refers to the unpredictability of the future and the occurrence of disruptions (Gleißner, 2011). The cause based understanding therefore is similar to some definitions of uncertainty. Instead we follow the effect based interpretation of risk (Hölscher, 1987) which puts the consequences of disruptions into focus and reflects to the hazards of not achieving project goals. According to that understanding risk describes an evaluation quantity providing information about the likelihood for damage as well as the expected impact of that incident (e.g. Conrow, 2003). Here damage must be interpreted as a loss caused by not achieving schedule, costs and quality objectives.

4.3 Scenario

In order to clarify the demand for a model based risk management approach, a scenario is presented reflecting the experience made within a German federal research and development project. The project aimed at providing innovative solutions for search and rescue robots operating in unstructured environments.

In the project the development of a snake-like robot was considered which offers a high degree of kinematic redundancy, enabling it to operate in collapsed buildings. The robot as well as its development process is complex with regard to the number and variety of system elements, development activities, involved disciplines and the strong interdependencies between those. An intense need for risk management was identified, as the project partners had to deal with uncertainties regarding imprecise target definitions, the technology in development and the design process itself. While in some cases uncertainties could be reduced by local changes of the technical solution or local modifications of the development process, the majority of cases called for macroscopic response, affecting interrelated project parts in a significant manner.

A representative example of uncertainty regarding the technical system is discussed in the following: The snake like motion concept of the robot was realized by four similar modules interconnected by joints, each offering five degrees of freedom. As part of the modular design approach, a special motor-unit was developed and implemented for each degree of freedom. When conducting a risk analysis we identified uncertainties regarding the performance characteristic of the actuator. No qualified decision could be made whether the torque provided by the joint would be sufficient to lift the robot's sensor head, as shown in Figure 4.1.

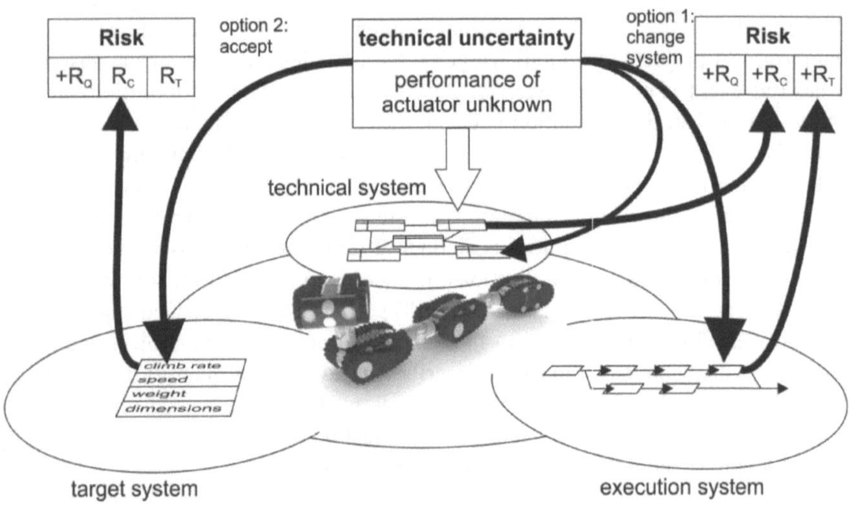

Figure 4.1. Heterogenic model of the system domains with uncertainty, response options and resulting risks

4.3.1 Treatment of Risks in the Presented Scenario

To deal with this uncertainty, several response options were considered that Gericke formally defines as *preventive*, *reactive* and *proactive* risk response (Gericke, 2011).

Preventive risk treatment
Preventive risk treatment aims on a reduction of risk by removing its causes. In the present example one could change the technical system by redesigning the motor-unit in order to allow for the integration of a more powerful motor. Changing the concerned part of the system will obviously reduce the uncertainty, but will also extend project duration (time risk R_T) and cost (cost risk R_C) due to the redesigning process. Moreover, the new motor-module will probably increase the system weight, consequently reducing the operating duration (quality risk R_Q). Finally, the changes to the technical system in turn will cause changes to other parts of the system, e. g. the chassis elements, resulting in additional quality, cost and time risks.

Reactive risk treatment
Reactive risk treatment addresses the impact of the risk and is applied not until the risk event has occurred. In the present example one could wait until tests with a physical prototype of the snake robot provide exact results. Changes at this time will probably result in broad schedule and budget overruns. One can also accept the risk of insufficient torque, conducting no changes at all. In that case the decision will result in a reduced quality of the product, but schedule and budget overruns can be avoided.

Proactive risk treatment
Proactive risk treatment also aims on a reduction of the effect of the risk, but risk treatment measures are selected before the risk occurs. In the example hardware-in-the-loop tests could be applied in order to acquire the necessary characteristics of the motor-module proactively. The engineering design of the test bench and conducting the tests will require additional project time and result in budget overruns but there is a chance that no negative impact on the quality occurs at all in case the original design proves valid.

4.3.2 Discussion of the Scenario

The presented example shows that a suitable modelling approach supporting the risk management in innovation projects has to integrate several domains, in particular the ones represented in the target system, the technical system and the engineering system and has to manage the dependencies in between. Moreover, the scenario demonstrates that an adequate modelling approach simultaneously assesses the effects of uncertainties and response strategies within all three dimensions of the *iron triangle*: *quality*, *costs* and *schedule*. In order to communicate uncertainties and risk management associated information between all stakeholders, the approach furthermore has to support the formalized description and assessment of uncertainties and potential response options.

4.4 Related Work

While, to the best of our knowledge, none of the approaches presented in literature complies with the outlined situation in product development satisfactorily, related work can be identified in the research areas of Quality and Change Management.

In the field of quality management primarily methods and models are provided to support the analysis of uncertainty effects in the dimension of quality. The well known Failure Modes and Effects Analysis (FMEA) e.g. aims at an early identification and formalized assessment of failures, taking into consideration the likelihood of occurrence (O), its significance (S) and probability of detection (D). Failures are prioritized by assigning the risk priority number, defined as the mathematical product of O, S and D. In comparison Fault Tree Analysis (FTA) and Event Tree Analysis (ETA) allow for a more detailed diagnosis. The underlying risk model of FTA is based on the principle of causality, expressing that each fault can be traced back to at least one cause. A set of lower level causes is defined that are connected to each other using Boolean logic. ETA inverts the principle of FTA and studies the effect of an initiating event on the system.

While System FMEA, FTA and ETA put quality aspects of the technical system into focus, approaches in the field of change management are provided that mainly concern the effects of changes in the dimensions of costs and schedule. The Design Structure Matrix (DSM) is widely used in order to investigate change propagation quantitatively. Clarkson *et al.* introduce the Change Prediction Method (CPM), using DSMs for tracing potential change propagation paths among the interconnected components of a technical system (Clarkson *et al.*, 2004). Chua and Hossain analyse the propagation of changes considering the development process and its interrelated design activities (Chua and Hossain, 2012). Smith and Eppinger present a model based on DSMs to simulate activity durations and probabilities for iteration (Smith and Eppinger, 1997). Beside such domain specific approaches, focusing either on the product or process domain, attempts are made to expand the analysis of change propagation across multiple domains. Koh *et al.* investigate the dependencies between requirements and components (Koh *et al.*, 2012). Tang *et al.* present a method linking entities in the product domain to the process and organization domain (Tang *et al.*, 2008). Ahmad *et al.* introduce a cross-domain approach to identify change propagation including the information domains of requirements, functions, components and the detail design process (Ahmad *et al.*, 2013).

These approaches have in common that they either focus on the impact of an uncertainty to quality aspects, or the effects of a change to schedule or costs. None of these approaches offers an integrated view that encompasses all three presented dimensions of risk. Moreover, uncertainties and the resulting response strategies are only modelled indirectly as attributes of the system elements and are not explicitly expressed.

4.5 The Model Based Risk Management Approach

The Model Based Risk Management Approach presented here consists of two parts: The Integrated System and Risk Management Model on the one hand which serves as the informational backbone of the approach, allowing for an explicit description of risk related aspects from a product, process and requirements point of view. The modelling process on the other hand describes the application of the model within the risk management process. Both parts are described in the following sections.

4.5.1 Integrated System and Risk Management Model

The *Integrated System and Risk Management Model* (*ISRM-Model*) provides the basic structure representing the information and relationship between all elements of the model. It is composed of Domain-Structure and Domain-Mapping Matrices, creating one integrated Multiple-Domain Matrix. The model itself consists of two parts, the *Target System, Technical System and Engineering System Model* (*TTE-System Model*) and the *Risk Management Model*, which each are represented by Multiple-Domain Matrices. Figure 4.2 shows the topography of the model.

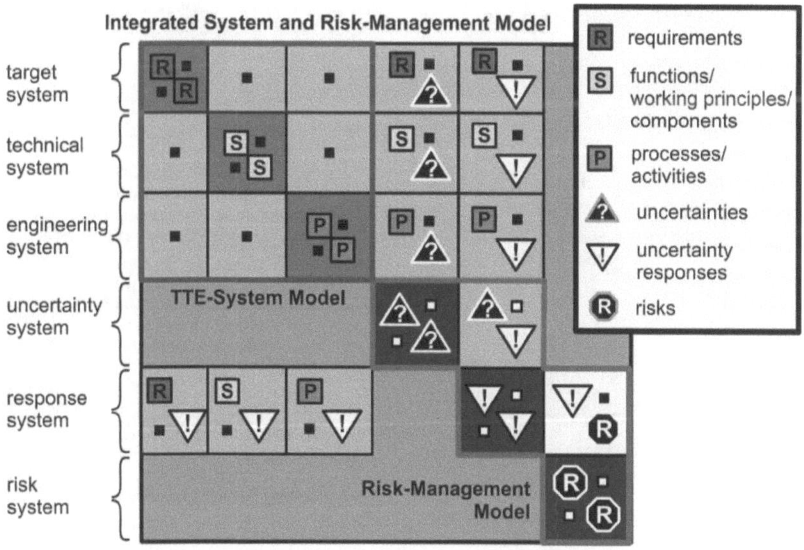

Figure 4.2. Structure of the Integrated System and Risk Management Model (ISRM-Model)

The *TTE-System Model* contains three domains, denoted as *target system*, *technical system* and *engineering system*. The *target system* is used to describe and structure all requirements while functions, working principles, components and their relationships are modelled in the *technical system*. The *engineering system* represents the development process and its activities, and details the information flow in between. Domain-Mapping Matrices are used to express cross-domain relations.

The *Risk Management Model* provides information about risk management associated aspects. The *uncertainty system* allows the formalized description of uncertainties and their assignment to related elements of the TTE-System Model. In the *response system* potential reactions to uncertainties are modelled. The modelling approach distinguishes different strategies for handling uncertainties which are discussed in the following section. For each response option associated elements are marked in the appropriate Domain-Mapping Matrices. Finally, the *risk system* holds information about the calculated risks that are caused by one or a group of response options.

4.5.2 Modelling Process

The modelling process describes the course of action when applying the ISRM-Model in the risk management process. The process we propose is a recursive procedure carried out in five stages. Figure 4.3 provides an overview of the procedure and the relevant model elements.

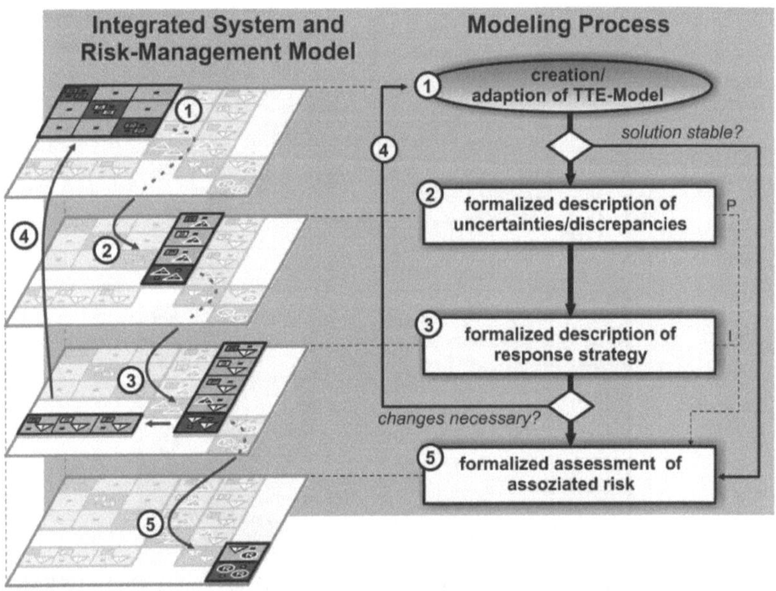

Figure 4.3. Modelling process

In the first stage the TTE-System Model is created or adopted to reflect the current status of the project. Uncertainties within the current situation are identified in stage two. These are described in the uncertainty system using standardized forms and linked to the affected sources. Thereafter a formalized description of potential response strategies is conducted in the response system. Uncertainties handled by the chosen response strategy are marked as well as the associated elements in the TTE-System. If changes are necessary, the process iteratively continues at stage one. When the solution is stable after the changes are carried out (i.e. there are no new uncertainties resulting from the changes) or there is no need for further changes, the process enters stage five where a formalized assessment of the risk associated to the response strategy is conducted. Risk is finally calculated as the product of likelihood and impact. Each risk is represented by a triplet of quality-risk, cost-risk and time-risk. This value is used to evaluate the response strategies developed to deal with a set of uncertainties.

The approach defines three basic types of response strategies: With "proactive action", changes to the TTE-System are incorporated immediately, regardless of the chance that the uncertainty might not occur (this usually leads to an over engineered solution while meeting or exceeding all requirements). The second strategy, "no action", represents the response option that the uncertainty is accepted and no response is carried out (this usually leads to quality risks). With the third strategy, "reactive action", the changes necessary to respond to an uncertainty are planned, but will only be executed when it is certain (i.e. as a result of a test) that the uncertain event occurs.

4.5.3 Application

To clarify the presented results, the proposed approach is applied to the example introduced in chapter three (Figure 4.4). The application scenario addresses the actuator unit as a standardized key subsystem of the mobile robot. The correct dimensioning of the motor with regard to its performance characteristics is questioned as a consequence of unknown friction forces and complex load profiles arising especially in unstructured terrain. Two different response options are shown in Figure 4.4 and compared in order to identify the best response solution matching the given situation.

Response option A applies the "no-action" strategy which results in limited mobility of the robot and ultimately in a loss in sales. As an alternative option B, a proactive change in the design and application of a more powerful motor is considered. This engineering change increases the weight of the system, affects the schedule (rework of P3) and costs.

For both response strategies, structure information is modelled using the provided matrix representation, while detailing descriptions of each element are conducted using standardized forms (simplified representation).

When stable solutions for both evaluated response options are achieved, the risk triplet is calculated (see Figure 4.3). For the final decision which response strategy is applied, a manual interpretation of the resulting risk has to take place, taking all company and marketing-strategic consideration into account as well as contract constrains.

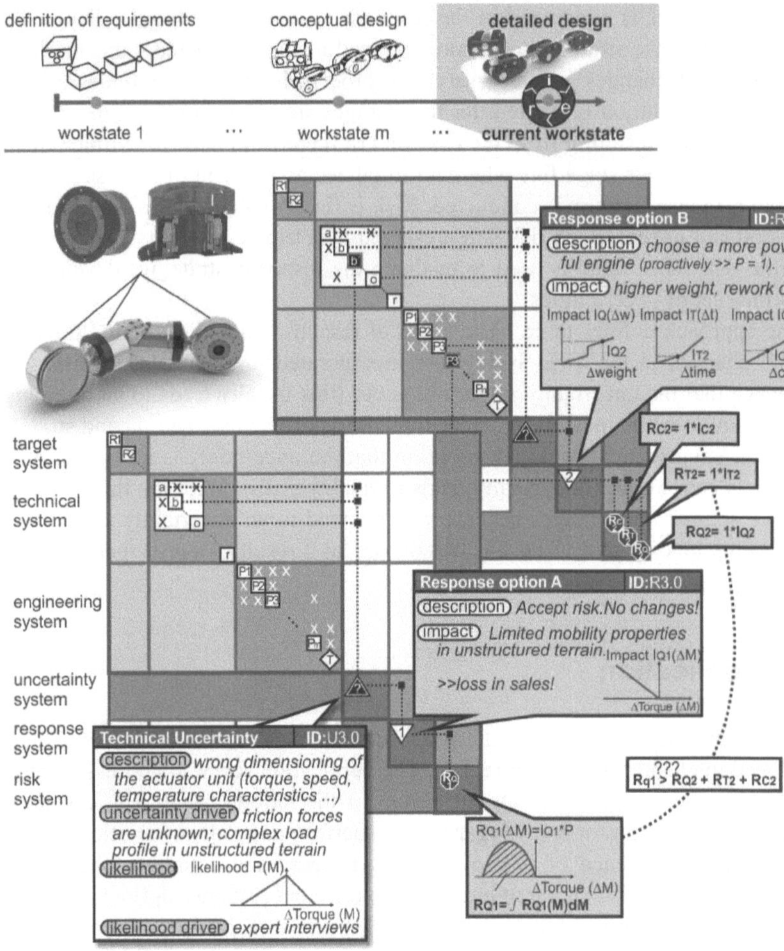

Figure 4.4. Model based risk management approach applied to the development of a search and rescue robot

4.6 Conclusions

Analysing risks early in the development process and treating them adequately with respect to effort and benefit is the key to effective risk management in new product development. This paper has presented the Model Based Risk Management approach, an integrated Multi-Domain approach to model risks and all aspects of system design explicitly in one consistent framework. The concept uses Multiple-Domain Matrices in order to integrate the product development point of view (represented by the domains target system, technical system and engineering system) and risk management (represented by the domains uncertainty system, response system and risk system). The approach can be applied to (1) model uncertainties and response options, (2) systematically evaluate those strategies and (3) support risk orientated decision processes in new product development.

The general applicability of the method could be demonstrated within an example. However, questions remain unanswered concerning the initial identification of uncertainties which is not supported by the approach up to now. Providing methodical support for uncertainty identification will therefore be subject to further research. We are planning to integrate established and well known methods, like Scenario Analysis or Delphi Method into the approach in order to establish a holistic framework. Furthermore, the demonstrated application of the method already indicates that practical usability is strongly related to the implementation of the method in a software tool. Prospective research therefore will consider the transfer of current research results into a software program.

4.7 References

Ahmad, N., Wynn, D. C., Clarkson, P. J. (2013) Change impact on a product and its redesign process: a tool for knowledge capture and reuse. In: Journal of Research in Engineering Design, Vol. 24, no. 3, p. 219-244

Bitz, H. (2000) Abgrenzung des Risiko-Frühwarnsystems i.e.S nach KonTraG zu einem umfas-senden Risiko-Managementsystem im betriebswirtschaftlichen Sinne. Betriebswirtschaftliche Forschung und Praxis, Vol. 52, no. 3, p. 231-241

Boehm, B. W. (1988) A Spiral Model of Software Development and Enhancement. Computer, Vol. 21, no. 5, p. 61-72

Browning, T. R. (1998) Modeling and Analyzing Cost, Schedule and Performance in Complex System Product Development. PhD thesis, Massachusetts Institute of Technology, Cambridge, Massachusetts

Chua, D. K. H., Hossain, M. A. (2012) Predicting change propagation and impact on design schedule due to external changes. In: IEEE Engineering Management, Vol. 59, No. 3, p. 483-493

Clarkson, P. J., Simons, C. S., Eckert, C. M. (2004) Predicting change propagation in complex design. In: Journal of Mechanical Design, Vol. 126, No. 5, p. 788-797

Conrow, E. H. (2003) Effective Risk Management: Some Keys to Success. 2nd edition, American Institute of Aeronautics and Astronautics, Virginia

Ehrlenspiel, K. (2007) Integrierte Produktentwicklung – Denkabläufe, Methodeneinsatz, Zusammenarbeit. 3. Auflage. Carl Hanser Verlag München Wien, München

Engelhardt, R., Wiebel, M., Eifler, T., Kloberdanz, H., Birkhofer, H., Bohn, A. (2011) UMEA – A Follow-Up to Analyse Uncertainties in Technical Systems. International Conference on Engineering Design, Technical University of Denmark

Ericson, A., Kastensson, A. (2011) Exploit and Explore: Two Ways of Categorizing Innovation Projects. In: Proceedings of the International Conference on Engineering Design 2011, Copenhagen, Denmark

Ferreira, V. V., Ogliari, A. (2005) Guidelines for a risk management methodology for product design, Product: Management & Development, Vol. 3, no. 1

Gericke, K. (2011) Enhancing Project Robustness: A Risk Management Perspective. Dissertation, Technische Universität Berlin, Berlin, Germany

Gleißner, W. (2001) Grundlagen des Risikomanagements im Unternehmen. 2. Auflage, Verlag Franz Vahlen GmbH, München

Hastings, D., McManus, H. (2004) A Framework for Understanding Uncertainty and its Mitigation and Exploitation in Complex Systems. In: 2004 Engineering Systems Symposium, Cambridge

Hölscher, R. (1987) Risikokosten-Management in Kreditinstituten – Ein integratives Modell zur Messung und ertragsorientierten Steuerung der bankbetrieblichen Erfolgsrisiken. In: Schriftenreihe des Institutes für Kreditwesen, Universität Münster

Koh, E. C. Y., Caldwell, N. H. M., Clarkson, P. J. (2012) A method to assess the effects of engineering change propagation. In: Journal of Research in Engineering Design, Vol. 23, no. 4, p. 329-351

Maurer, M. (2007) Structural Awareness in Complex Product Design. Dr. Hut Verlag, München

Smith, R. P., Eppinger, S. D. (1997) A Predictive Model of Sequential Iteration in Engineering Design. In: Management Science, Vol. 43, No. 8, p. 1104-1120

Smith, P. G., Merritt, G. M. (2002) Proactive Risk Management – Controlling Uncertainty in Product Development. Productivity Press, New York, USA

Tang, D., Xu, R., He, R. (2008) Design structure matrix-based engineering change management for product development. In: International Journal of Internet Manufacturing and Services, Vol. 1, no. 3, p. 231-245

VDI (2004) VDI-Richtlinie 2206: Entwicklungsmethodik für mechatronische Systeme. Beuth Verlag, Berlin

Verworn, B. (2005) Die frühen Phasen der Produktentwicklung: Eine empirische Analyse in der Mess-, Steuer- und Regelungstechnik. Dt. Univ.-Verlag

Weck, O., Eckert, C., Clarkson, J. (2007) A classification of uncertainty for early product and system design. In: Proceedings of the International Conference on Engineering Design 2007, Paris

Chapter 5
Estimation of Risk Increase Caused by Parallelisation of Product Development Processes

N. Szélig, M. Schabacker and S. Vajna

5.1 Introduction

In the ever-changing technological and economic environment, time plays an increasingly important role. Only those products can be successful and competitive, which come to market at the right time. Several studies have shown that in the development of series products a delayed market launch is more expensive than an increase of the development costs (Melboldt *et al.*, 2012).

It has become the focal point to develop product development processes and methods that are capable of getting the product to market within the desired time frame. In addition to the solution of technical problems, other aspects are given weight in these methods: the product development process organisation, the time span, resources, capacity, planning of costs and information processes, quality assurance of the development process, and the conscious handling of risks connected to innovation (Rick, 2007).

For effective product development, it is necessary to identify, to observe, and to control all processes and activities. In order to shorten the development process, it has to be optimised. The resulting increase in risk is estimated to allow conscious handling or to minimise it through the help of other factors.

After modelling a process with the appropriate method (e.g. BPMN (Freund and Rücker, 2010), Container modelling (Schabacker and Wohlbold, 2002), DSM (Rick, 2007)) follows the optimisation of the process (Schabacker *et al.,* 2013) (Szélig *et al.*, 2012).

After topology optimisation (Vajna *et al.*, 2005), (Vajna *et al.*, 2011) the process elements are rearranged to be either fully parallel or linear (possibly alternative or iterative). In a next step, the remaining linear elements can be further optimised with Simultaneous Engineering, where these elements are parallelised to a certain degree (Vajna *et al.*, 2005). This will be further discussed in Chapter 5.2.

M. Schabacker et al. (eds.), *Modelling and Management of Engineering Processes*, DOI 10.1007/978-3-662-44009-4_5

5.2 Parallelisation of Linear Process Elements

This paper deals only with the optimisation stages Simultaneous Engineering and Concurrent Engineering. For these stages it has to be determined what percentage of a process element needs to be completed in order to start the next process elements. This can be achieved reasonably with the use of the documents to be created, such as CAD models, technical drawings, and product documentation (Schabacker *et al.*, 2010).

The **degree of fulfilment** is the minimal percentage of a document that has to be completed to provide the minimal amount of information necessary to begin the follow-up document (early transfer of partial results).

The degree of fulfilment needed for parallelising process elements is thus measured by the partial completion of documents. Therefore, document types will be defined which can be seen in Figure 5.1.

Depending on the process and the company, the extent of overlapping of process elements and thus the degree of fulfilment for parallelising process elements may vary. For simultaneous elements a lower limit for the time advance must be introduced, with which the earlier element completes before the later element (called *minimum time advance - m*), to ensure that the later element, which depends on the information of the earlier element, has enough time to run. Some possible degrees of fulfilment (orange) and the minimum time advance (blue) values for the documents are represented in Figure 5.1. Surveys can determine the percentage for both.

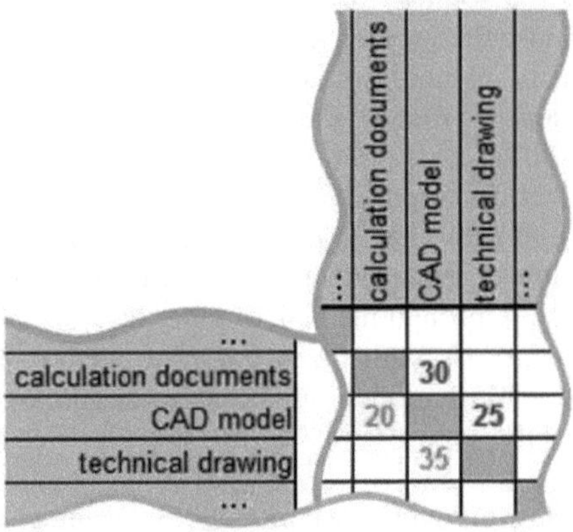

Figure 5.1. Document types with possible degrees of fulfilment and minimum time advance in percentages

If multiple documents are created in a single process element and a premature beginning of a document within a process element is possible, it is useful to divide the process element into sub-process elements (Concurrent Engineering), where each sub-process element contains exactly one document, and therefore multiple commissioners can work on different documents and sub-process elements in parallel.

The time overlap of normally sequential workflows thus provides a bonus time and/or a shortened processing time, respectively. As soon as sufficient information is gathered in a workflow, the next workflow is started in parallel. This sometimes leads to more work, because the element cannot always be operated with the final level of information, but the basis for work may change at any time.

For sequential process elements a time overlap is possible. A process element can be initiated before the previous item has been completed. The processing of the element can start with a certain amount of information delivered by the predecessor. The further data are supplied continuously. The predecessor must be ended earlier than the current element, so that all information can be adopted.

In the representation, the arrows that do not begin at the end of the element but at a certain point (with given percentage) indicate that, at this degree of fulfilment, overlapping is possible (Figure 5.2). These arrows lead to the beginning of the next element. Additional arrows from the end of a predecessor to a point in the current element indicate where no further proceeding is possible without the final data.

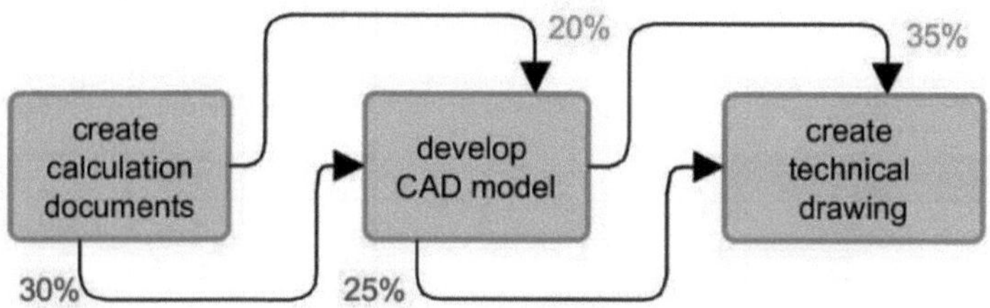

Figure 5.2. Representation of a simultaneous case

Degree of parallelisation

To characterise the parallelisation a variable is needed. This is the degree of parallelisation. Its maximum value is calculated from the degree of fulfilment and the minimum time advance.

The lower limit of the degree of fulfilment provides the highest parallelisation, along with the highest risk. In this case, it may happen that the element needs to be divided into several parts, to ensure that the minimum termination condition is satisfied. If partial elements are undesirable, the degree of parallelisation is obtained by a comparison of the weighted difference between the degree of fulfilment and element length (100%) with the weighted difference between the minimum termination and the length of the next element. The smaller of these two differences is the degree of parallelisation of two elements. The degree of parallelisation of the overall process is the sum of the individual parallelisation. Standardisation is already taking place through the individual weightings, the sum of which is always exactly one.

Example: In a sub-process, three documents - a *calculation document, CAD model, and the technical drawing* - are created. Of course, a *calculation document* doesn't need to be 100% completed in order to derive the *CAD model* or to begin with the *technical drawing.* Perhaps the *technical drawing* can be performed in parallel with the *CAD model.* Furthermore, the project manager will be able to select the best possible qualification profile for working on each of the three documents separately. Instead of assigning a design engineer to work on all three documents, the project manager can give the technical drawing to a draftsman, which under certain circumstances may lead to lower process costs, due to the lower hourly rate (Figure 5.3 and Eqn. (5.1)).

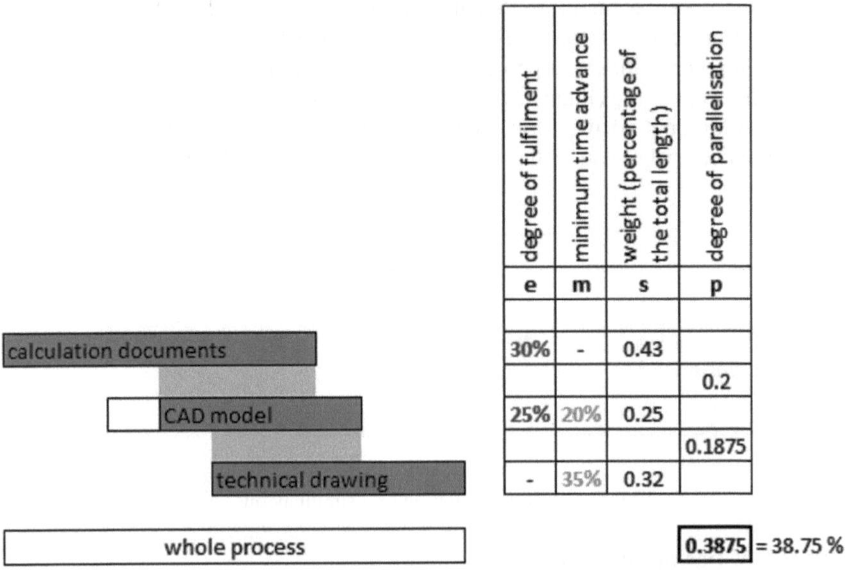

Figure 5.3. Sample data for the calculation of the degree of parallelisation

$$p_{process} = \sum_{i=1}^{n} \min\left[s_i(1-e_i); s_{i+1}(1-m_{i+1})\right] \tag{5.1}$$

$$where \sum_{i=1}^{n} s_i = 1$$

$$and \sum_{i=1}^{n-1} p_i = p_{process}$$

with n: number of the elements in the process

In this example, the derivation of the *CAD model* can be started at the earliest when the *calculation document* is completed to 30% (degree of fulfilment – e_1). After the completion of the *calculation document*, the *CAD model* takes at least

20% of the time (minimum time advance - m_2) to get ready. Only in special cases may these values be the limits according to the predetermined percentages, so that smaller overlaps must be taken. For this, the weight is calculated (percentage of the total length) of the individual elements, $s_1 = 0.43$ and $s_2 = 0.25$. For instance the largest possible overlap of element *calculation document* is s_1 (1-e_1) = 0.43 (1-0.3) = 0.301 and that of the element *CAD model* is s_2 (1-m_2) = 0.25 (1-0.2) = 0.2. The minimum value is 0.2, which is the degree of parallelisation p_1.

To be able to use this parallelisation method, the following boundary conditions must be kept:

- It is assumed that only the relation to the directly preceding element needs to be considered. In this example, Figure 5.3, the predecessors of the element *technical drawing* are the elements *CAD model* and *calculation document*. The *technical drawing* can only begin when both the *CAD model* and the *calculation document* have achieved a certain degree of fulfilment. Nevertheless, only the degree of fulfilment of the *CAD model* is considered and it is assumed that this already includes the degree of fulfilment of the *calculation document* regarding the *technical drawing*.
- Furthermore, the splitting of elements is excluded. A premature start of an element is conceivable if it is interrupted as soon as new information is needed. However, here is assumed that a once begun element is executed continuously until the end, and cannot be broken into several parts.

The lengths of the individual elements are weighted so that the sum of all the single p_i is the degree of parallelisation for the entire process. If there is no parallelisation, the value is 0%, the elements run sequentially. If all the elements run at the same time, this value becomes 100%.

5.3 Risk Increases When Parallelising Within the Process/Project

In order to obtain a common understanding of some of the terms used in this paper, they are predefined as follows:

A **process** consists of interrelated activities or sub-processes for performing a task. The number of activities is not limited in their length and duration. The compounds of the activities or sub-processes are not rigid. Thereby a sub-process is the subset of a process and also a set of activities or other sub-processes ((Hammer and Champy, 1993), (Hammer and Champy, 1995), (Freisleben, 2001), (Schabacker, 2001)).

A **project** is a living process (or several connected ones), in which boundary conditions are defined and which is always unique (DIN 69901, 2009).

Risk in general is the occurrence of a negatively considered event (Seiler, 1995).

The definition of risk considered in this work is a time related risk for which the loss of time is interpreted as a damage (no other types of damage are considered).

„Risk is the negative deviation between the actual times (durations) of the intended process/sub-process/element and the desired times (durations)."

In principle, any technical system can be implemented. Ideally, there is at least one technical solution for every technical problem and thus no technical risk. The risk arises when a limited cost and time budget is provided to implement the technical system. Except of some very specific research projects an unlimited budget is hardly conceivable and completely excluded for projects within the free economy. Thus, schedule-, cost-, and quality- risks arise, which can be represented as project risks, due to economic constraints (Hänggi, 1996).

The project management triangle (Atkinson, 1999) shows certain constraints of a project. Traditionally these constraints are listed as:

- time: duration and schedule
- cost: personnel costs and other resources cost (here resource)
- scope: content and quality of the results.

In this triangle each side represents one of the constraints. It cannot be changed one of the sides without affecting the others. Further, to each constrain also belongs a risk. Including these risks into the graphical representation one arrives at a tetrahedron (see Figure 5.4). In the tetrahedron, it is possible to fix a size, and consider only the other three sizes.

Figure 5.4. Tetrahedron of the constraints

During the optimisation of the process, the expected value for the total duration of the process is getting smaller while the parallelisation increases. It is apparent that the larger the parallelisation, the greater the risk that the process with the given resource cannot be completed in the required quality at the scheduled time.

Meanwhile, the consciously acceptance of risks under application of risk management techniques became an important driver in market economy.

In the tetrahedron of the constraints a balance must be achieved in which the target can be reached under acceptable risk and the requirements on resources, time and quality.

The risk is that the project is not completed in the scheduled time with the intended resources in the expected quality. An important property is that the risks can be calculated, while under conditions of uncertainty no expected value can be specified (Knight, 1921), that could be used by the tetrahedron.

If there is any evidence that one or more variables do not match the original plan, corrective actions should be set. If the originally desired time is reduced by parallelisation (Simultaneous Engineering), the risk increases that the new deadline needs to be moved or personnel costs must be increased. If these corrective actions are omitted, the project target is highly endangered and additional risks occur.

Improved triangle for risk estimate

In order to manage the risks, a slice of the tetrahedron of the constraints – an improved triangle – is used to estimate the changes (Figure 5.5). In this work the quality is fixed, i.e. no changes in quality are made and the results have to be of equal quality.

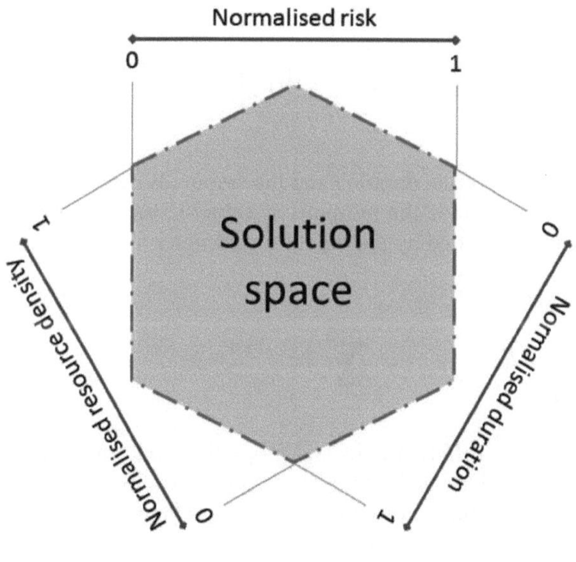

Figure 5.5. Improved triangle

The improved triangle assumes the given factors (time, resources and risk density) on three normalised axes in two-dimensional space and shows the dependencies of the three factors.

With simple algebraic skills it can be demonstrated that the solution space is a hexagon. The projections on the three axes of a point (x values) have the same constant sum for all points within this hexagon (Eqn. (5.2)). x_R, x_Q and x_Z are the projection values of the risk, of the resource and of the time. Thus, with two known values, the third is always expressible.

$$x_Z + x_Q + x_R = c \Leftrightarrow x_R = c - (x_Z + x_Q) \tag{5.2}$$

Since all three factors build different functions, and the relation between them is not linear, functions in a general polynomial form are placed on the three axes. These functions characterise the variables belonging to the individual factors. $Z(x_Z)$ is the duration or at the same time the inverse of the degree of parallelisation, $Q(x_Q)$ is the resource density and $R(x_R)$ the risk (Eqn. (5.3), (5.4) and (5.5)). z, q and r are the coefficient of the time, resource and risk polynomials.

$$Z(x_Z) = \sum_{i=0}^{n} z_i x_Z^{\,i} \quad \forall\ x_Z \in [0,1], \tag{5.3}$$

$$Q(x_Q) = \sum_{i=0}^{n} q_i x_Q^{\,i} \quad \forall\ x_Q \in [0,1], \tag{5.4}$$

$$R(x_R) = \sum_{i=0}^{n} r_i x_R^{\,i} \quad \forall\ x_R \in [0,1] \tag{5.5}$$

From the functions of the duration and the resources density, the risk is expressible. Equation 5.6 describes the normalised value of the risk, where the function parameter x_R was substituted by the function parameter x_Z and x_Q according to Eqn. (5.3).

$$R(x_R) = R(x_Z, x_Q) = \sum_{1=0}^{n} r_i (c - x_Z - x_Q)^i \tag{5.6}$$

The variables have different units and properties, so normalisation is necessary. This is done by projecting the function values on an axis with a range of 0 to 1. It is important that the functions are unique, i.e. every function value f(x) may occur only for a single function parameter (x).

The functions can and must fit the specific product development processes. The character is similar, but the exact shapes differ significantly. The detailing of the functions of the individual factors follows below.

5.3.2 Parallelisation and Duration

The duration function is a linear function (1st degree polynomial), because the duration always changes linearly with all changes of the degree of parallelisation. Only the slope of this function may vary.

In more complex processes it may occur that the degree of parallelisation is rising, but the duration is not reduced below a certain point. However, in this case, the risk to the overall processes doesn't rise any further either. Therefore, it is sufficient to calculate with the linear function of the duration. This function is shown in Figure 5.6.

$$Z(x_Z) = z_0 + z_1 x_Z \tag{5.7}$$

The largest recorded value of the function (at $x_Z = 1$ on the normalised scale) is the total time duration of the process, without parallelisation.

The smallest value (at $x_Z = 0$ on the normalised scale) is the time duration at full parallelisation (z_0), which is the longest element length. This is a pure theoretical minimum value. It will never be reached in a real process, where the elements depend on each other and need information from each other. Otherwise, the process could take place in parallel at once, without concurrent engineering.

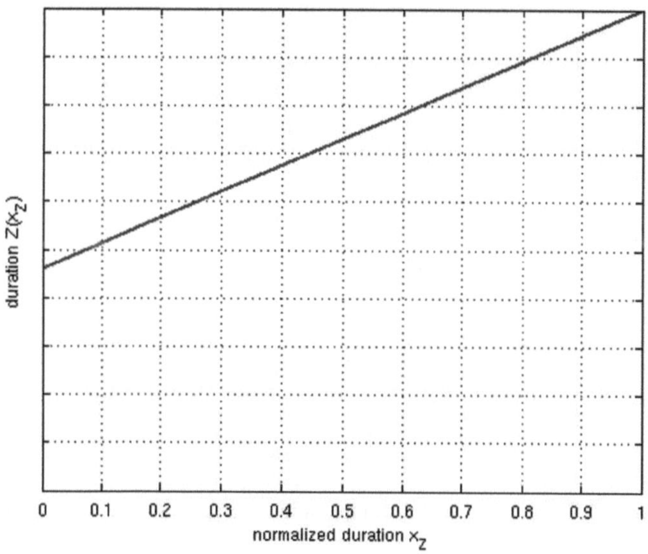

Figure 5.6. General shape of the graph of the duration function

5.3.3 Resource Density

Resource density shows the distribution of resources in the individual work elements. It may be different in each element; the average resource density characterises the process.

The function of resource density is approximated very well with a polynomial of third degree (Eqn. (5.8)). The graph of the function is monotonically increasing with an inflection point at $x_Q = 0.5$. Here it changes from convex to concave.

$$Q(x_Q) = q_0 + q_1 x_Q + q_2 x_Q^2 + q_3 x_Q^3 \tag{5.8}$$

The smallest value belonging to the process is located on the function at $x_Q = 0$, which is in the resource density of the parallelisation free process. This is the sum of all the products of element lengths (durations) with their resource requirements divided by the total time period of the process. The highest value is set to $x_Q = 1$ on the normalised scale and it is as large as the largest parallelisation needs. It is the sum of the resource requirements for all elements (Figure 5.7).

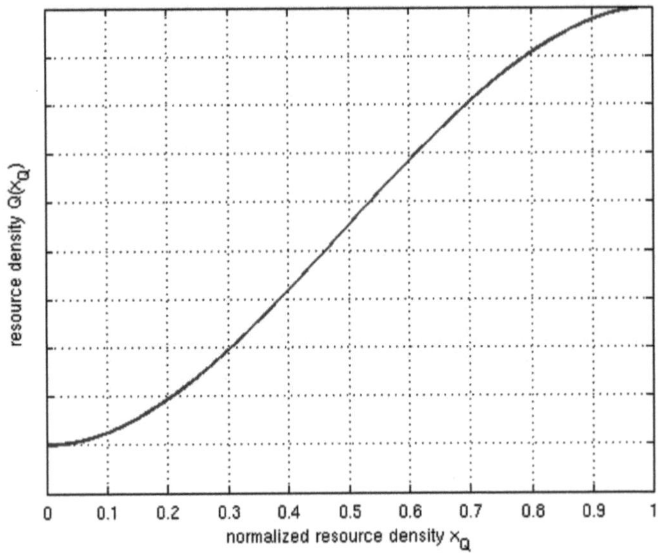

Figure 5.7. General shape of the graph of the average resource density function

5.3.4 Risk

Experience shows that the risk can be represented well with a second-order polynomial (Eqn. (5.9)). The graph of the function is monotonically increasing and convex.

Its values have to be calculated from time duration and resource density. There is also the possibility of forming risk categories and assigning the values to them. Then, the usual risk determination can be used similarly. Figure 5.8 shows a general risk function graph from which the trend of the function is clearly visible.

$$R(x_R) = r_0 + r_1 x_R + r_2 x_R^{\ 2} \tag{5.9}$$

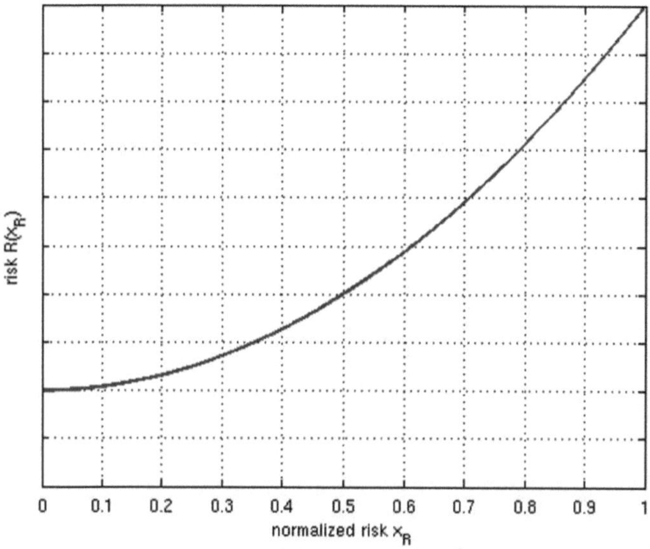

Figure 5.8. General shape of the graph of the risk

The Eqn. (5.10) is derived from Eqn. (5.6) for n = 2. It provides the possibility to determine the risk directly out of the time duration and the resource density.

$$R(x_R) = R(x_z, x_Q) = r_0 + r_1(c - x_Z - x_Q)$$
$$+ r_2(c - x_Z - x_Q)^2 \tag{5.10}$$

By integrating the graphs into the triangle (Figure 5.9) one can estimate the risk from the two other factors: the duration and resource density. In Figure 5.9 two cases are shown: the first with 28% parallelisation, the other with 50% parallelisation and accordingly significant differences in the duration. With a well selected resources density a compensation is possible such that, despite the simultaneous short process flow, the risk (that it does not proceed as intended) is not much greater.

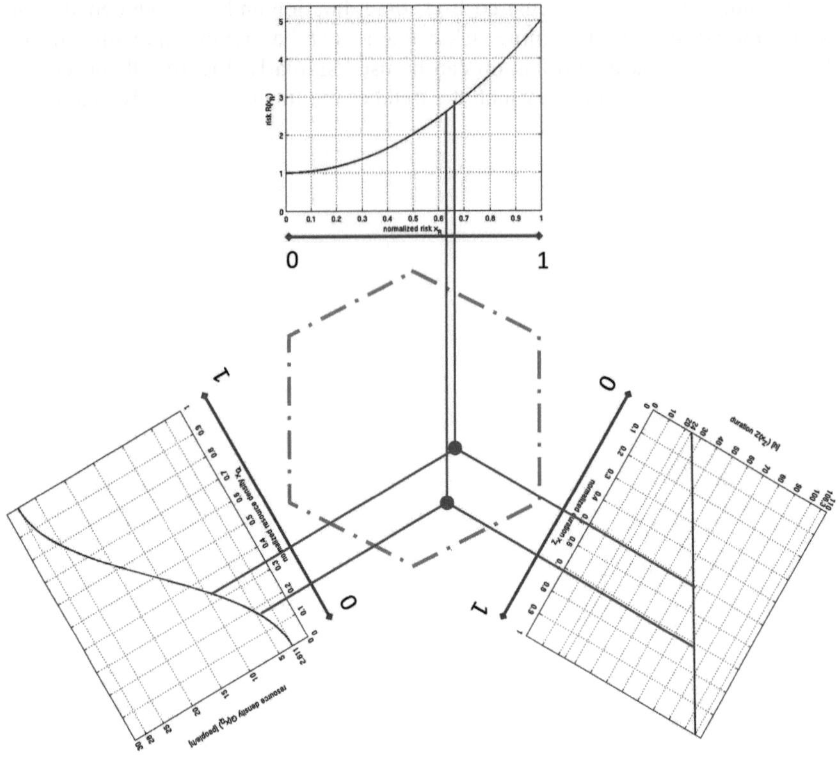

Figure 5.9. Risk estimate with improved triangle

5.4 Summary and Outlook

Companies applying the optimisation approaches discussed above will be able to perform better and to run more efficient product development projects. The above mentioned assessment and optimisation approaches allow the product development cycle times to be shortened, thereby reducing the cost of product development and improving the utilisation of project participants in on-going product development projects. It is during this optimisation that the risk to the project increases. With the elaborated improved triangle method the risk can be estimated and countermeasures can be initiated, e.g. an increase of the resource density. In future work the improved triangle could be stocked with functions whereby the correlations between the factors become more detailed and fit even better into the processes. The next step is to automate the derivation of the functions and thus the calculation of the risk and vice versa, the possible combinations of the other factors to a desired level of risk.

5.5 References

Atkinson, R. (1999) Project management: cost, time and quality, two best guesses and a phenomenon, its time to accept other success criteria, International Journal of Project Management, Vol. 17 (6) pp. 337–342

DIN 69901 (2009) Projektmanagement; Projektmanagementsysteme, 2009-01

Freund J., Rücker, B. (2010) Praxishandbuch BPMN 2.0, 2. Aktualisierte Auflage, Carl Hanser Verlag München Wien, pp. 8-17, 28

Freisleben, D. (2001) Gestaltung und Optimierung von Produktentwicklungsprozessen mit einem wissensbasierten Vorgehensmodell, Buchreihe Integrierte Produktentwicklung, Band 2, Otto-von-Guericke-Universität Magdeburg

Hammer, M., Champy, J. (1995) Business Reengineering: Die Radikalkur für das Unternehmen, 5. Auflage, Campus-Verlag, Frankfurt/Main

Hammer, M., Champy, J. (1993) Reengineering the Corporation, Harperbusiness, New York

Hänggi, R. (1996) Risikomanagement und Simultaneous Engineering; St. Gallen, Univ.; Dissertation

Knight, F. (1921) Risk, uncertainty, and profit. Boston, USA: Hart, Schaffner & Marx, Houghton Mifflin Company

Melboldt, M., Matthiesen, S., Lohmeyer, Q. (2012) "The Dilema of Managing Iterations in Time-to-market Development Processes". In: *Proceedings of the 2nd International Workshop on Modelling and Management Engineering Processes (MMEP 2012)*, Cambridge, United Kingdom, pp. 85-97

Rick, T. (2007) "Gépipari terméktervezési folyamatok erőforrás és költségszempontú optimálása a termékstruktúra figyelembevételével". PhD Thesis, Budapesti Műszaki és Gazdaságtudományi Egyetem, Gépszerkezettani Intézet, Budapest, pp. 27-32

Schabacker, M. (2001) Bewertung der Nutzen neuer Technologien in der Produktentwicklung. Buchreihe Integrierte Produktentwicklung (ed Vajna, S.), Band 1, Otto-von-Guericke-Universität Magdeburg

Schabacker, M. (2010) "PDM-Systeme: Wirtschaftlichkeitsberechnung von PDM/PLM-Investitionen". *PLM-Jahrbuch 2011*, Hoppenstedt Publishing GmbH, pp. 62-65

Schabacker, M., Szélig, N., Vajna, S. (2013) "Suitable Methods for Process Modeling and Process Optimization". *Proceedings of 2013 ICED: International Conference on Engineering Design,* 19-22 August 20013, Sungkyunkwan University, Seoul, Korea, ICED13/176

Schabacker, M., Wohlbold, L. (2002) "Benefit Evaluation of EDM/PDM Systems". *Proceedings of the 8th International Conference on Concurrent Enterprising* 17-19 June 2002, Rome, Italy, pp. 413-420

Seiler, H. (1995) Rechtsfragen technischer Risiken - Aufsätze zum Thema Risikorecht. Zürich: vdf

Szélig, N., Schabacker, M., Vajna, S. (2012) „Determing the Degree of Parallelisation of Processes in a Tri-process-modelling-tool". *Proceedings of the 2nd International Workshop on Modelling and Management Engineering Processes (MMEP 2012)*, Cambridge, United Kingdom, pp. 119-129

Vajna, S., Clement, St., Jordan, A., Bercsey, T. (2005) "The Autogenetic Design Theory: An Evolutionary View of the Design Process". *Journal of Engineering Design* 16, 4, pp. 423-440

Vajna, S., Kittel, K., Bercsey, T. (2011) "The Autogenetic Design Theory-Product development as an Analogy to Biological Evolution". *The Future of Design Methodology*, Springer Verlag, pp. 169-179

Chapter 6

Emerging Telemedicine Analysis of Future Teledermatology Application in France

T. A. Duong, R. Farel and J. Le Cardinal

6.1 Introduction

World health organization (WHO) defines Telemedicine (TM) as the use of IT-technologies to provide clinical healthcare at distance. This past decade, scientific production provided a great number of publications focusing on TM feasibility studies and medical applications (Henderson *et al.,* 2013). Its indisputable interest was raised to provide an equal access to care for isolated population or population with chronic diseases. Increase demand for healthcare, patient triage efficiency in the healthcare system, decrease number of physicians are also challenges for implementing telehealth. Faces to a mature technology, many countries search how to successfully implement TM and to move pilot studies or small-scale trials to mainstream deployment. TM was previously described as a complex multi-stakeholders innovation and an immature service with various management processes (Barlow *et al,*. 2006). Local policy or local healthcare demands increase difficulties to generate a generic system and its sustainable implementation. TM is an organizational and social challenge for patients, healthcare professionals, and decision makers, a technological challenge for service industries or IT-industries. Barriers to its implementation are linked to organizational and cultural context as well as sustainable funding and economical model.

Teledermatology is the dermatological application of TM for skin disorders, with two described methods real time, i.e. Live interactive LI using videoconferencing system or non real time i.e. store-and-forward (SF) using clinical data and images transfer. TD medical interest comparing its diagnosis accuracy or management to conventional care is well established (Warshaw *et al.*, 2011). In France, TM legal framework was recently fixed by an ordinance in 2010. For a population of sixty million inhabitants, the decrease number of specialists, the increase ageing population and the unequal access to dermatologic healthcare with a specialist density varying from 1 dermatologist per 20,000 inhabitants to 1 per 40,0000 inhabitants (Syndicat des dermatologues.org) justifies the implementation of TD system to decrease time to consult delay or improve medical access. For all stake-

M. Schabacker et al. (eds.), *Modelling and Management of Engineering Processes*, DOI 10.1007/978-3-662-44009-4_6

holders TM and TD don't aim at replacing conventional care but to improving medical access or healthcare organization, with limited resources. Being a practitioner dermatologist and component of this system, the motivation behind this research was to understand the benefit of implementing a TD organization regarding the reference model of the conventional care.

6.2 Background

6.2.1 Healthcare as a Complex System

Healthcare devices or organizations development are submitted to several visible e.g. local policy, local healthcare political objectives, restrict law, device approval of regulatory agencies such as FDA and invisible constraints e.g. inappropriate users, lack of users, not user centered, reliability, cost of maintenance (Medina *et al.*, 2013).

Healthcare and telehealth were previously described as complex system regarding the multiplicity of their elements, interactions and relationship (Schindler *et al.*, 2007).

Systemic approach to design complex organizational structures was described in SCOS'D (Systemics for Complex Organisational Systems' Design) by Bocquet *et al.* This method enables stakeholders' identification and expectations, process creation, value (ethical, scientific, environmental, societal) creation and evaluation (Schindler *et al.*, 2007). For healthcare organizations, challenges include an efficient and optimized resource allocation. Healthcare organization management includes various scientific areas e.g. economy, management, sociology, optimization and risk analysis. In a generic approach, the identification of stakeholders, preexisting organization, value creation and key performance indicators or variables are useful to implement statistic model to assess or simulate an organization efficiciency efficacy and risk (Aktaş *et al.*, 2007; van der Geer *et al.*, 2009; Basole *et al.*, 2013; Robinson *et al.*, 2012).

6.2.2 Efficiency in Healthcare

Efficiency is determined by an optimized management of resources and costs (Peck *et al.*, 2010). Healthcare resources scarcity and rising costs place efficiency as a major concern for decision makers (Robinson *et al.*, 2012). To improve it, optimization of management resources and of patient care delivery processes i.e. hospital processes, medical processes and non medical processes are necessary (Devaraj *et al.,* 2013). Using processes indicators, measures of quality of care/ access (patients or professionals interviews) or metric mesures of processes efficiency contribute to determine efficiency. Processes indicators are selected and developped on the basis of mesurability, validity and controllability. In healthcare they include patient outcome (feedback evaluation), quality of life, physical or psychological test, or process indicators i.e process solving, procedural (van der Geer *et al.,* 2009). Balance scorecard method was also used to determine performance indicators in healthcare (Grigoroudis *et al.*, 2012). This approach is widely used for strategic planning and management in business industries or non-profit/governemental organizations. From four different perspectives, i.e. financial, customer, internal business and leaning and growth, this method gives a balanced view of organizational performance adding non-financial strategic indicators to financial indicators. For healthcare organizations, performance indicators could be for the financial perspectives: net profit margin or inventory turnover; for the "customer" i.e patients perspectives: average waiting time, average duration of hospitalisation, satisfactions; for the internal business: employee absenteeism, surplus inventory; for learning perspectives: resources allocations for IT, budget used for purchase of new technology, number of new projects with other organisations. In our perspective, measureable key perfomance indicators could be process indicators, sucha as time and cost. Theses process indicotors could ad minima integrate the quality of care and financial outcomes.

6.2.3 Design for Healthcare

Clarkson defines Design as a structured process to identify problems and to develop and to evaluate user-focused solutions. It was successfully used to transform products, services, systems and even entire organizations. (Clarkson, 2004)

Design is used in the field of service, product, management or organization. It aimed to identify the system requirements, the input/ output of each phases of a development process. In healthcare, design can either be used for technological or device development as well as to guarantee the safety and quality of a medical process, to identify tasks or determine a conceptual model (Clarkson *et al.*, 2004; Alexander and Clarkson, 2002).

6.2.4 Process Modelling in Healthcare

Diagrams offer a rapid mapping, a simple model to capture key concepts or the reality (Aurisicchio and Bracewell, 2013). Process modelling is a key activity to set the problem or for effective quality improvement. Healthcare processes include medical treatment processes and generic organizational processes (Jalote-Parmar and Badke-Schaub, 2008). They are highly multidisplinary, highly dynamics, and complex and successfully used to determine system boundaries or possibilities.

In healthcare, several methods were identified regarding the problem to state or the component to model: stakeholder diagrams, information diagram, process content diagrams, flow-chart, swim lane activity diagram, state transition diagram, communication diagram and data-flow diagram (Jun *et al.,* 2009). To optimize a procedure or activity, process modelling is useful to identify key indicators, variables, actors or activities. Breaking down healthcare organizational model processes from macro to micro processes was previously used with system of metrics to quantify capacities, utilizations and process flows to improve tasks and resource efficiency (de Mast *et al.*, 2011).

6.3 Methodology

Implementing telemedicine is converting a new technology to improve a preexisting organization and network model.

Using DRM methodology, this paper describes the stage of "problem understanding" i.e Descriptive study I. To determine the potential benefit of implementing TD, we aimed at understanding the conventional care process (Figure 6.1).

Our approach included both the systems description (stakeholders) and the generation of scenario of conventional care process for a skin problem with 3 key actors: the patient, the general practioner (GP), the dermatologist and the reference model of conventional care activity. Avoiding unecessary in-person visit to the specialist, the future development of TD will impact sub-activities or sub-processes of the conventional care process for those three actors justifying our focus. We first described and modeled our conventional care process activities. Then two process indicators time and cost were selected as key variables for further simulation to analyze the conventional care process for a skin problem and the impact of TD store and forward process implementation. This approach should help all stakeholders to display their role and to plan TD usage regarding their initial investment or expectations. It should also allow the quantification of the CC process and the determination of its improvement when modifying some activites.

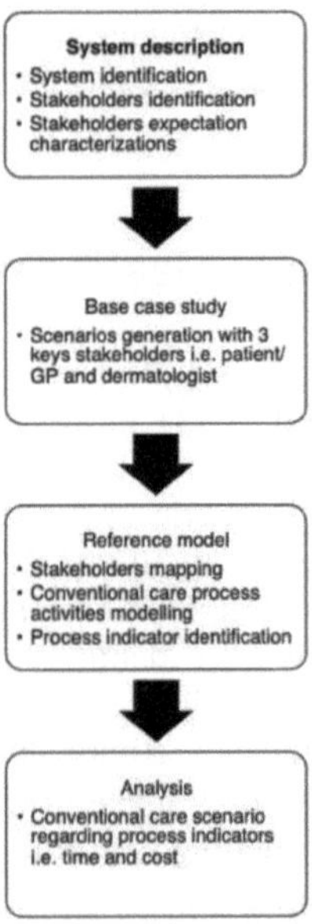

Figure 6.1. Descriptive study I of the conventional care

6.4 Results

6.4.1 Conventional Care Process Stakeholders

The conventional care (CC) is the existing system where any patient will enter for care. In our case the CC is the in person visit to the either the GP or the dermatologist when having a skin problem. Behind the three main actors, the conventional care also includes various stakeholders such as State, insurances, hospitals, National physician unions etc. Their interactions with CC might be classified following their role i.e. fiancial investors, assessors, other systems, IT-professionals, requested and requesting physician and patients (Figure 6.2).

Examples are: insurances reiumburse patients visit, State fix the number of student in med-school and healthcare campaigns etc.

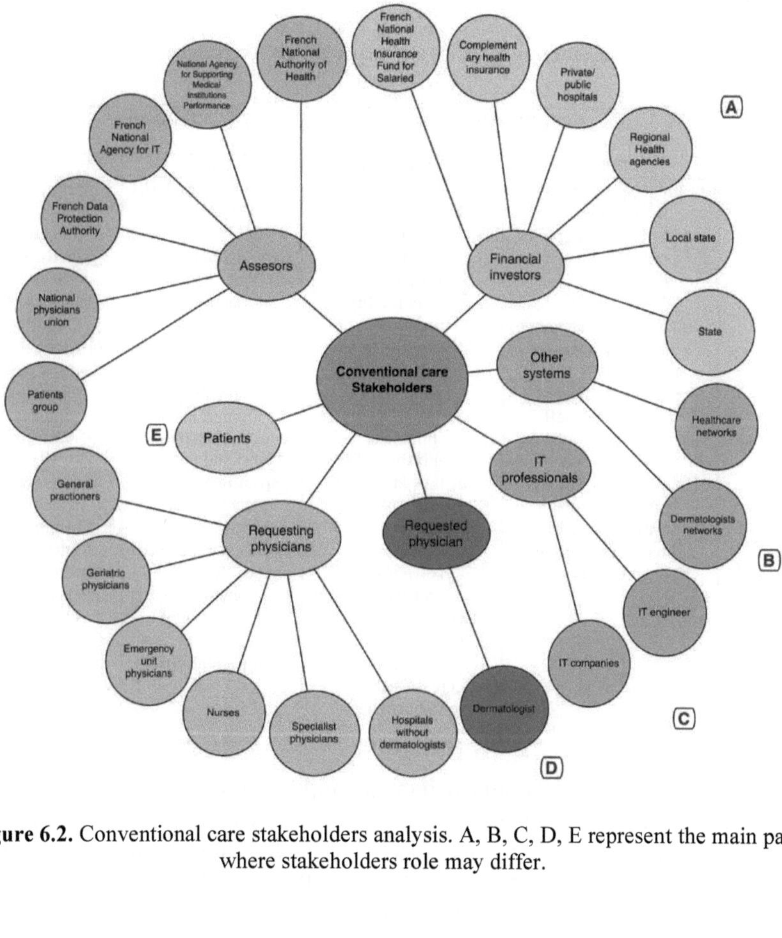

Figure 6.2. Conventional care stakeholders analysis. A, B, C, D, E represent the main part where stakeholders role may differ.

Their role may be predominant or not in TD, ex-IT professionals are necessary to develop or implement TD tools (C) while their role is optional in the conventional care, e.g. TD may focuss on given care to patients with disabilities, acute or chronic skin disorders or leaving far from any dermatologist (E).

Regarding their function and description, some stakeholders' role and expectations are described in Table 6.1, for both CC and TD (HAS, 2013; DGOS, 2008). For financial investors such as the state, additional investment to finance TD may contribute to increase an equal medical access or create new jobs while for the French national insurance funds of salaried workers it may decrease the cost of transportation reimbursement. To add, the great majority of financial investors for both the CC and TD are public organizations. Interestingly, State is an uncommon financial investor that can fix the law, the contraints or the use of this new system. In the expectation of new job creation or the perspective of research program investment, their expectations could be directly link to care delivery improvement but also the development of new tools or devices to answer to TD technological needs.

Table 6.1. Financial stakeholder roles and expectations for both conventional care and teledermatology

Stake-holders	Conventional care		Teledermatology	
	Role	Expectations	Role	Expectations
FINANCIAL INVESTORS	Finance physicians training Fix the healthcare policy Fix public-health orientations Fix the law Finance public hospitals Finance medical care center Finance medical or paramedical staff Identify and fix medical organization Reimburse the conventional care process Pay sick leave Finance prevention program	Prevention program Decrease infant Mortality Decrease dermatological diseases morbidity and mortality Equal access to care Increase working population Publications Research Influence Organization of the medical landscape Ensure an equal territorial medical access Receive insurers contributions Decrease morbidity costs associated to illness Decrease sick leave reimbursement	Fix and order reports from public-health agencies Vote the law regarding Telehealth Finance investment projects Research programs Finance local investment Invest and develop specific networks Finance material investment Create and Finance local telehealth organization Finance pilot program Pilot reimbursement	Contribution to an equal access of care Increase states attractiveness Ensure social equity New jobs creation Research program publications Influence Increase innovation Offer new services to members Opening new markets Control medical costs and reimbursement Decrease cost of medical transportation Regulation of medical fare

For dermatologists, TD implementation could optmize their consultation time and select patients seen by primary care physicians and requiring an in-person visit. For patients, TD may be accessible under certain situation to avoid unecessary travel. For our three main actors in the conventional care process, this analysis highlights that a new process integration would require additional fees, or optimize both physician and patient time or also increase physicians knowledge (Table 6.2).

Table 6.2. Key actors: Primary care physician, patient, and dermatologist roles and expectations for both conventional care and teledermatology

Stake-holders	Conventional care		Teledermatology	
	Role	Expectations	Role	Expectations
REQUESTING PHYSICIANS	Provide medical care	Improve dermatological diseases management Ask for specialist consult Getting paid for the consult	Provide medical care Integrate a novel process	Improve dermatological diseases management Continuous medical education Specialist network Improve access to a second opinion Integrate a benefit from a new medical activity
REQUESTED PHYSICIANS	Provide dermatological care None	Getting paid for medi-cal care	Provide dermatologi-cal care Assist dermatologist or requesting physi-cians Integrate a new job	Getting paid Time optimization Patients triage Communication network Getting salary
PATIENTS	Request care	Being treated or orien-tated in the healthcare system	Integrate a novel process Pay for a novel process in term of clinical research or extra-fees	Being treated or orientated in the healthcare system decrease delay to consult ensure medical access

This stakeholder analysis highlights the heterogenity of their expectations and points the difficulty to design a system satisfying all of them. At least based on the clinical usage it should integrate user needs and constraints. In a context of ressource scarcity or political will to implement or develop an innovative system, the success of this new system implementation relies on its capacitIcy to meet stakeholders expectations.

6.4.2 Conventional Care Scenarios for a Patient With Skin Disorders

From the patient with skin disorders, 16 scenarios including the 3 main actors were generated using a graphical tree model. These scenarios include the possibility for a patient to visit a GP in a hospital when serious or severe conditions. Tree scenario highlights all the medical options in the conventional care process for a patient having a skin disorders. For a patient having a skin problem, this graphical diagram displays that a problem resolution could include two GP consult processes or one GP and one specialist consult process (Figure 6.3).

An example: A patient with skin disorders will go to a family practice to visit a GP, who can solve the problem or send the patient to get additional exam or specialist consult. In case of a severe or urgent case, GP may send the patient to the hospital or to a private specialist practice.

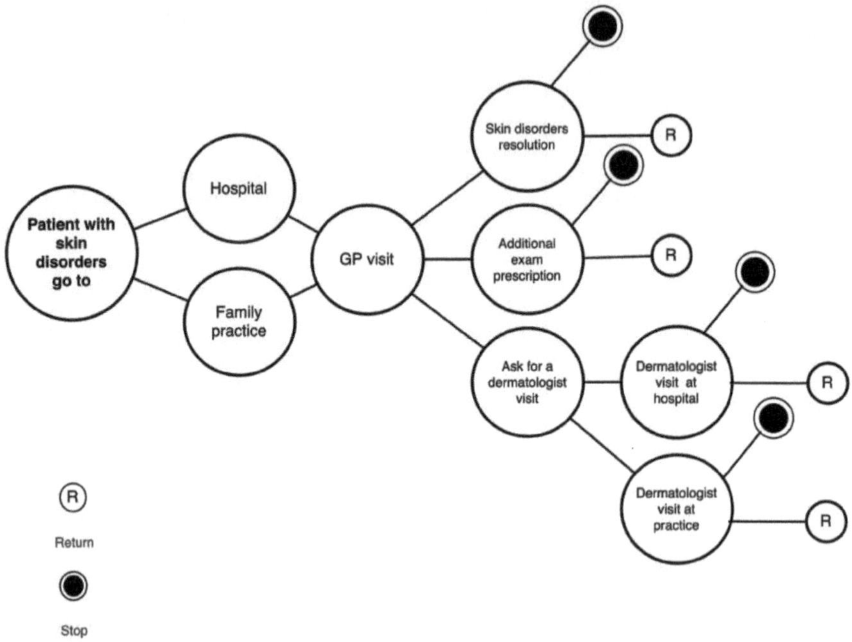

Figure 6.3. Tree scenario in the conventional care process of a patient with a skin problem

6.4.3 Reference Model Conventional Care Process

Because of the care journey process, patients will first, in most of the cases, schedule a visit to their GP. He may solve the skin problem, prescribe additional exam or send the patient to the specialist. Modelling the process activities of CC when skin disorders highlight the necessity for the patient to go to and go back from GP office to get a consult. This also identifies process indicators for each activity of the CC process such as time or lead time for transportation, consultation, scheduling visit, and stop working. As seen on the process flow Figure 6.4, some of these activities may be immediate or non immediate e.g. to schedule an appointment to the GP or to the dermatologist.

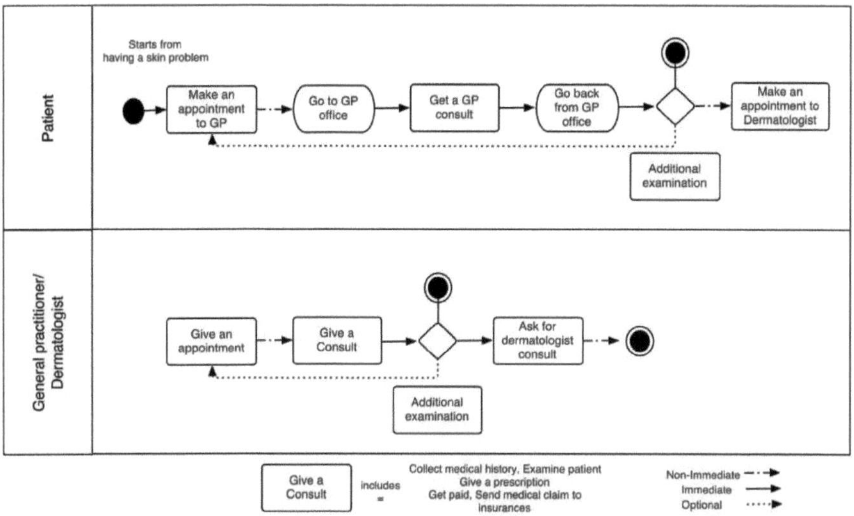

Figure 6.4. Conventional care, activity modeling for a patient with a skin problem

Further, to illustrate the advantages or inconvenience of the CC process, we modeled the activities and determined the required procedures for each actvities i.e. make an appointement to the dermatologist and go to the dermatologist. For the patient, TD system could skip inconveniences linked to CC i.e. finding a dermatologist, organizing his working schedule (Figure 6.5). It could influence or decrease the lead time to get a consult, the waiting time and the transportation time. If necessary, the patient could get his specialist consult from a direct interaction between his GP and the dermatologist. Using TD for a specialist medical advice or consult would also suppress activities for the patient and add new ones to the GP (Figure 6.6). This innovation would require new tasks integration for GP, such as taking picture, loading them or filling specific form while in the CC, he would only have to draw up a letter (Figure 6.7). This process modeling approach displays that TD could be time saving for both the patient and the dermatologist who do not see patient in-person but not for GP. At this point it is a crucial to determine the direct benefit or value for GP to gurantee a sustainable use of this new system. It could either be financial i.e additional fees for this new service or scientific value i.e continuous medical education, or skin disease management improvement.

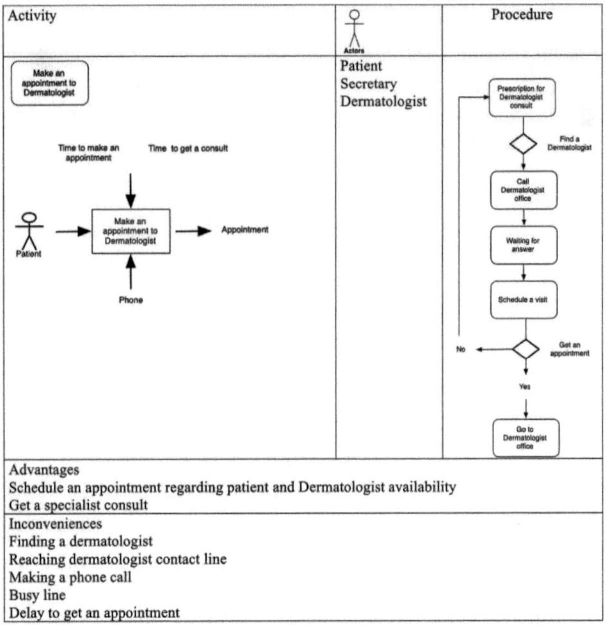

Figure 6.5. Conventional care, procedure to make an appointement with the dermatologist

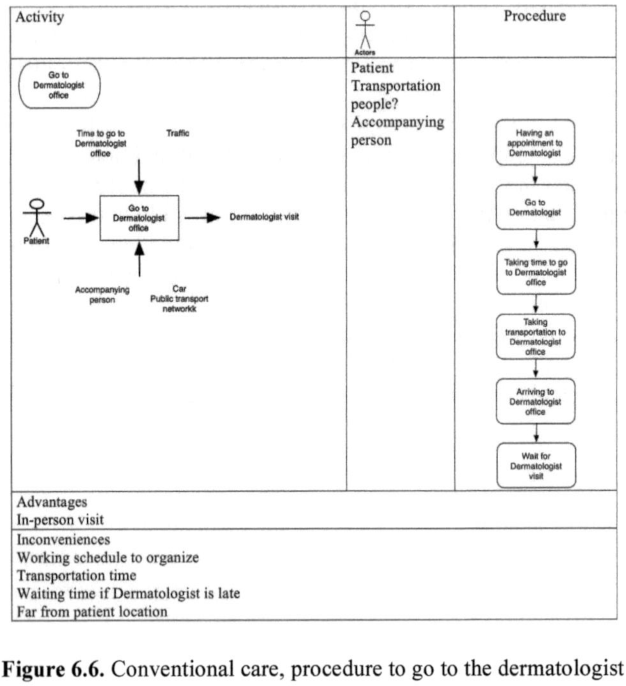

Figure 6.6. Conventional care, procedure to go to the dermatologist

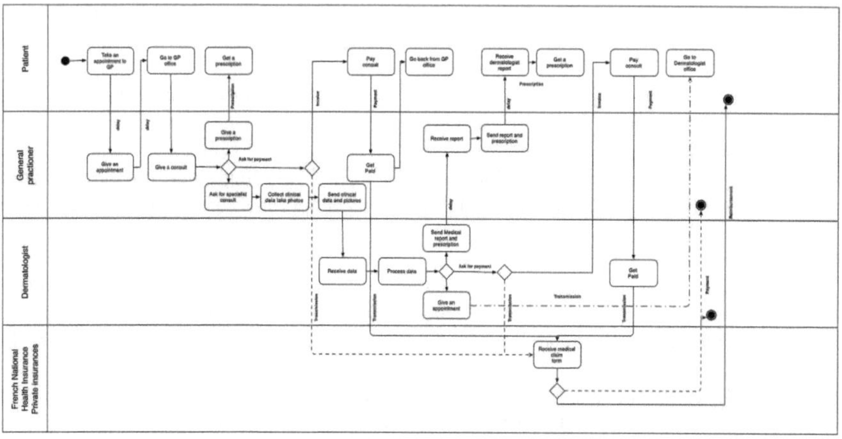

Figure 6.7. An example of store and forward Teledermatology process

In this preliminary work, we described and modeled a conventional skin disorders care process. Subprocesses and activities analysis pointed out its advantages and inconveniences. Time and costs were identified as mesurable process indicators to perform an analysis of this system efficiency. In the case study we gathered numerical data of these process indicators to quantify the conventional care process. Some further studies will include a analytic model regarding all scenarios for those 3 actors, and their activity model when using TD. This approach breaking macro processes to micro processes enables the quantification and the comparison of both CC and TD processes. This work should help us to point the real benefit of implementing TD regarding stakeholders expectations and its real impact on the CC optimization.

6.5 References

Aktaş, E., Ülengin, F. & Önsel Şahin, Ş., (2007) A decision support system to improve the efficiency of resource allocation in healthcare management. *Socio-Economic Planning Sciences*, 41(2), pp.130-146

Alexander, K. & Clarkson, P.J., (2002) A validation model for the medical devices industry. *Journal of Engineering Design*, 13(3), pp.197-204

Aurisicchio, M. & Bracewell, R., (2013) Capturing an integrated design information space with a diagram-based approach. *Journal of Engineering Design*, 24(6), pp.397-428

Barlow, J., Bayer, S. & Curry, R., (2006) Implementing complex innovations in fluid multi-stakeholder environments: Experiences of 'telecare'. *Technovation*, 26(3), pp.396-406

Basole, R.C., Bodner, D.A. & Rouse, W.B., (2013) Healthcare management through organizational simulation. *Decision Support Systems*, 55(2), pp.552-563

Clarkson, P.J. *et al.*, (2004) Design for patient safety: A review of the effectiveness of design in the UK health service. *Journal of Engineering Design*, 15(2), pp.123-140

Devaraj, S., Ow, T.T. & Kohli, R., (2013) Examining the Impact of Information Technology and Patient Flow on Healthcare Performance: A Theory of Swift and Even Flow (TSEF)

Perspective. *Journal of Operations Management.* Available at: http://www.sciencedirect.com/science/article/pii/S0272696313000272 [Consulté le octobre 18, 2013]

Van der Geer, E., van Tuijl, H.F.J.M. & Rutte, C.G., (2009) Performance management in healthcare: Performance indicator development, task uncertainty, and types of performance indicators. *Social Science & Medicine*, 69(10), pp.1523-1530

Grigoroudis, E., Orfanoudaki, E. & Zopounidis, C., (2012) Strategic performance measurement in a healthcare organisation: A multiple criteria approach based on balanced scorecard. *Omega*, 40(1), pp.104-119

Haute Autorité de Santé - Efficience de la télémédecine : état des lieux de la littérature internationale et cadre d'évaluation. Available at: http://www.has-sante.fr/portail/jcms/c_1622477/fr/efficience-de-la-telemedecine-etat-des-lieux-de-la-litterature-internationale-et-cadre-devaluation?xtmc=&xtcr=2 [Consulté le septembre 17, 2013a]

Henderson, C. *et al.*, (2013) Cost effectiveness of telehealth for patients with long term conditions (Whole Systems Demonstrator telehealth questionnaire study): nested economic evaluation in a pragmatic, cluster randomised controlled trial. *BMJ (Clinical research ed.)*, 346, p. 1035 f

Jalote-Parmar, A. & Badke-Schaub, P., (2008) Workflow Integration Matrix: a framework to support the development of surgical information systems. *Design Studies*, 29(4), pp. 338-368

Jun, G.T. *et al.*, (2009) Health care process modelling: which method when? *International Journal for Quality in Health Care*, 21(3), pp.214–224

De Mast, J. et al., (2011) Process improvement in healthcare: overall resource efficiency. *Quality and Reliability Engineering International*, 27(8), pp.1095–1106

Medina, L.A., Kremer, G.E.O. & Wysk, R.A., (2013) Supporting medical device development: a standard product design process model. *Journal of Engineering Design*, 24(2), pp.83-119

Peck, J., Nightingale, D. & Kim, S.-G., (2010) Axiomatic approach for efficient healthcare system design and optimization. *CIRP Annals - Manufacturing Technology*, 59(1), pp.469-472

Rapport_final_Telemedecine.pdf. Available at: http://www.sante.gouv.fr/IMG/pdf/Rapport_final_Telemedecine.pdf [Consulté le avril 29, 2013b]

Robinson, S. et al., (2012) Priority-setting and rationing in healthcare: Evidence from the English experience. *Social Science & Medicine*, 75(12), pp.2386-2393

Schindler, A., Bocquet, J.-C. & Dudezert, A., (2007) Systemic approach as a multi-criteria design method: healthcare R&D centre application. In *Proceedings of International Conference on Engineering Design.* Available at: http://hal.archives-ouvertes.fr/hal-00158077/ [Consulté le mai 27, 2013]

Syndicat des dermatologues. *Syndicat des dermatologues.* Available at: http://www.syndicatdermatos.org

Warshaw, E.M. et al., (2011) Teledermatology for diagnosis and management of skin conditions: a systematic review. *Journal of the American Academy of Dermatology*, 64(4), pp.759-772

Chapter 7
Evaluation of Collaborative Tools Throughout the Design Process Using a Quantitative Rating of CAD Model Modification

D. Fleche, J. B. Bluntzer, M. Mahdjoub and J. C. Sagot

7.1 Introduction

In the current industrial context marked by globalisation, the product design process can be achieved by various stakeholders who collaborate efficiently to offer innovative products onto the market (Couto *et al.*, 2006). Such a collaboration can take place in co-located or distributed meetings where decisions about the design project are made. This is a challenge, as each one of them comprises different goals, experiences and sometimes even cultures. Thus, various collaborative tools are used to optimise communication and ease collaboration among these actors. The use of collaborative tools allows product representations to be employed, and for dialogue and information exchange among stakeholders to be facilitated. This paper deals with the evaluation of the impact of using collaborative tools on project data modification. More precisely, this evaluation is based on the definition of the CAD model modification rate. In this paper, we reviewed the collaborative tools used during the design process and their current methods of evaluation. After having outlined the current limits of these evaluations, we proposed an approach based on the development of the project data. This approach is a calculation of the modification rate of the project data. This is a first step of an overall evaluation of the collaboration. In this paper, we propose a calculative approach of a modification rate indicator, and this is applied on a generic application.

M. Schabacker et al. (eds.), *Modelling and Management of Engineering Processes*, DOI 10.1007/978-3-662-44009-4_7

7.2 State of the Art

7.2.1 Collaborative Tools Used During the Product Design Process

The first category of collaborative tools used during the design process is that which allows the employment of product representations (e.g. drawing, CAD model, mock-up). The development of the design process has been defined by various authors throughout the 90s (French, 1998), (Pahl and Beitz, 1996), (Pugh, 1991). More specifically, we referred to the Pahl and Beitz approach, which today is a reference in much research work. Indeed, they have described the design process and divided it into four phases.

In the first phase, the "planning and clarifying the task" step, collaborative tools are mainly used to gather the stakeholders so that they can define the product design requirements. For instance, most collaborative tools used during this early phase of the design process are digital libraries (Wodehouse and Ion, 2010). The various actors involved must collect information and implement their knowledge, but there is little collaboration between them. This capitalisation at the early stages of the design process increases product productivity by reducing the level of engineering activity (Bluntzer *et al.*, 2011).

In the second phase, the "conceptual design" step, the project actors must share their knowledge and reach a decision to define one agreed solution (Ostrosi *et al.*, 2012). Contrary to phase one, there are a lot of exchanges of expertise between the actors and collaboration is important. The corresponding collaborative tools take different forms depending on the actors' requirement (e.g. interactive floor to generate an idea (van Dijk and Vos, 2011), interactive table or hybrid ideation space to generate forms (Dorta *et al.*, 2011), (Hartmann *et al.*, 2010)).

In the third phase, the "embodiment design" step, the actors define the concept specifically. In this phase, each actor implements his or her area of expertise on the product design and must consider each other's expertise. For example, virtual reality tools enable the actors to come together on the product design (Bennes *et al.*, 2012). Collaboration is important and there is a high risk of misinterpretation.

In the last phase, the "detail design" step, the product is finalised.

To summarise, during every phase of the design process, there are many different collaborations between the actors involved in the design project. Interaction between the various actors of the design project vary all throughout the design project. In order to respect the time constraints given by the client, collaboration among actors must be optimised, mainly by choosing the correct collaborative tool.

7.2.2 Evaluation of Collaborative Tools

Previous authors highlight that during the product design process, collaborative tools are developed and evaluated for a specific use in a specific situation (Antunes *et al.*, 2012). More precisely, it has been demonstrated that tools are evaluated through interviews and recordings of the project review, for instance, with virtual reality technology (Al-Khatib *et al.*, 2013), or with the study of actions and verbalisation (Defays *et al.*, 2012). Most of the time, the usability criterion is based on the observation of the interactions among the design project's actors, and on the analysis of subjectively-completed questionnaires. Moreover, studies demonstrated that collaborative tools can be evaluated through the quantity, quality, novelty and variety of their results (Geyer *et al.*, 2011). These kinds of qualitative evaluations give information about the usefulness of the collaborative tool. Hence, the evaluation methods are based on two criteria: usability and results. However, with this validation method, we cannot declare if the decisions made with these collaborative tools will be consistent all throughout the design process. Therefore, we propose an evaluation of the collaborative tools based on the product design process, through a new and additional criterion, called data development (Formula 7.1).

$$\begin{aligned}&\text{Suitability of the collaborative tool} =\\&\text{f (Exchanges, Results, Data Development)}\end{aligned} \quad (7.1)$$

In order to represent the product design process, we based our evaluation on the Product Lifecycle Management (PLM) approach. The PLM approach appeared at the end of the 90s and was created to manage every tool related to the design process (Stark, 2011). This management of every tool, related to the design process, includes the recording, storage and reuse of every piece of data from the project. The development of this data and its links represents the development of the whole product, all throughout the design process. In every phase of the design process, the data evolves through several states to reach a reliable state (Saint-Marc *et al.*, 2004). Our working hypothesis is that the use of a collaborative tool has an impact on the data development of the project. We propose that the evaluation of the collaborative tools should be based on three criteria: results, exchanges and data development (Figure 7.1). Indeed, with this approach, we assume that the development of the project data can be evaluated through the use of various collaborative tools (A, B and C). The suitability of one tool will then be linked to the previous tools used.

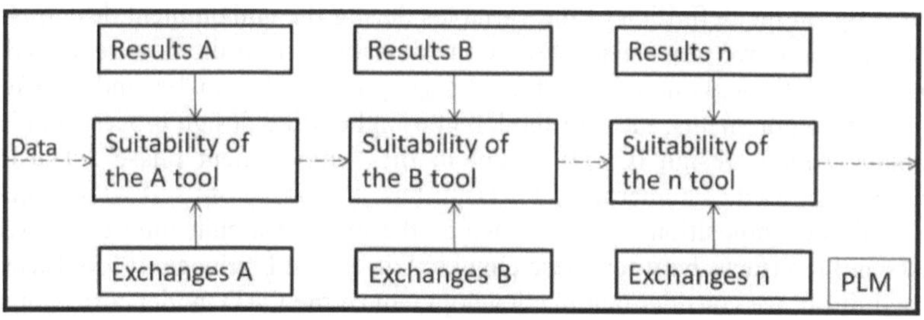

Figure 7.1. Evaluation of the collaborative tools

7.3 Proposed Approach

Through this article, we introduce a new approach to evaluate the use of collaborative tools based on the product design process data. In the text, the development of the product design process is mainly based on the study of the management of change propagation (Clarkson et al., 2004). Change propagation is defined as the "process by which a change to one part or element of an existing system configuration or design results in one or more additional changes to the system, when those changes would not have otherwise been required" (Giffin *et al.*, 2009). This definition describes the interconnection between the different product parts of the design project and the model of change propagation. Pasqual and De Weck (2011) introduced the notion of a multilayer network model in order to observe every modification during the product design process. The multilayer network is made up of three types of layers: the product layer, the change layer and the social layer.

- The product layer corresponds to the product or system being designed. It is made up of every component of the project, for instance the hardware or software components. Also, the technical interfaces can be represented, for instance the physical connections or information.
- The change layer corresponds to the change propagation of the process. In this layer, the change requests and every relationship's propagation through the change are represented. The change may result from "any modification to the form, fit and/or function of the product as a whole or in part" (Jarratt *et al.*, 2011). These engineering changes can result from emergent or initiated changes.
- The social layer corresponds to the organisation. It includes every person involved in the design project and their relationship amongst individuals and groups. With this layer, it is possible to understand every role the actors play and the impact of their decisions.

The definition of every layer and their links allows for the understanding of the design process and the impact of the changes on the product. They represent the various viewpoints of the project.

We focused on the development of the data generated on the product layer and specifically on the definition of the variables during the embodiment design phase. For this purpose, we propose to observe the development of the data generated and employed all throughout the product design process. It can be under different forms (e.g. requirements, ideas or models) throughout the design process and is the Computer Aided Design (CAD) model in the "embodiment phase". The CAD model is composed of numerous sub-assemblies, composed of itself with numerous parts. This decomposition is also known as the product architecture composed of chunks and interfaces between these chunks (Ulrich and Eppinger, 1995). Based on this definition, we consider that the development of the CAD model corresponds to the development of every sub-assembly and model parts. This development is impacted by the decisions made during the actors' collaborations.

Thus, our hypothesis is that the impact of these decisions on the development of the design process allows for evaluating the use of collaborative tools. This impact will highlight the fact of a common decision without containing any statement about the quality of the decision. Indeed, the optimisation of the design process corresponds to the reduction of design iterations at a micro level. We consider that the usefulness of a collaborative tool can be evaluated according to its impact on design process iterations. In order to test our hypothesis, we consider that the link between the use of a given collaborative tool and the CAD model is represented through its modifications.

The aim of the paper is to propose an approach to evaluate the modification rate performed on a CAD model.

7.4 Proposed Modification Rate

7.4.1 Modification Rate of the CAD Model

To define the calculation of the CAD model modification rate, we have observed an industrial case on how the CAD model is modified during the design phase. The industrial case in question is a design project made by one mechanical engineer with seven years of experience within the IRTES-SeT laboratory of the University of Technology of Belfort-Montbéliard, working with the client and some subcontractors. In this industrial project, the mechanical engineer created various versions of a CAD model according to the development of his objectives and various design meetings. His objectives represent the different goals that he has to reach in line with the overall aim of the project. The design meetings occur during the project when the client or the subcontractors have to make decisions regarding the project. During these meetings, collaborative tools are used as seen in section 7.2.1. Thus, we observed that all throughout the design project, the CAD modification type can be divided into three types of modification: the creation of a new part or sub-assembly, the removal or the modification of an existing part or sub-assembly. Moreover, the modification of a part has a different impact as regards the significance of the part on the system. If we observe a CAD part, we can define two variables representing the impact of this part on the system: the **functional definition of the part**, variable α, and **the impact on the environment**, variable β. These two variables define the significance of the CAD part on the system and can be represented as a percentage.

Thus, we assume that the modification rate directly depends on the significance of the parts and on the type of modification (Formula 7.2).

$$\text{Modification rate} = f(\text{Significance rate}, \text{Type of modification}) \qquad (7.2)$$

In this article, we focus on the mathematical definition of the significance rate of the CAD parts and its application in practice.

7.4.2 Functional Definition of the Part, Variable α

As seen in section 7.4.1, the variable α corresponds to the functional requirements and the demands of the customer. It is the link between the CAD part (*p*) and the flexibility of its functional requirements (*f*), named *Lfp*. Thus, a modification carried out on a part linked with major functions of the product will strongly impact the design process. Inversely proportional, a modification done on a part linked with minor functions of the product will not have a strong impact on the design process. In order to represent connections between various components like parts, tasks or people, the Domain Mapping Matrix (DMM) representation is used (Steward, 1981). Moreover, in a design project, the functional requirements have various significance as regards the flexibility of its functions, also called the value engineering method (Cross, 2000). The weight of each requirement *Wf* can be obtained by determining the flexibility (F) of the function criteria (NF EN 1325-1, 1996). This standard breaks down every functional requirement in order to define every criteria related to the function. These notions have also been defined by Pahl and Beitz (1996) in the identification and classification of the requirements. Four levels are also defined: the demand (flexibility 0, F0) which has to be respected and the wishes in a major (flexibility 1, F1), medium (flexibility 2, F2) or minor significance (flexibility 3, F3).

Thus, in this calculation, we used a multiplying factor for these four criteria. The distribution of the weight of the factor is based on a proportional relationship, where the flexibility of the criterion is a demand and has to be respected and the multiplication factor is equal to 1 (F0). The other levels of flexibility are respectively equal to 0.8 (F1), 0.5 (F2) and 0.2 (F3).

Consequently, the calculation of the functional definition of the part is the ratio between the sum of the links, *Lpf,* and the sum of the weight of the function, named *Wf* (Formula 7.3).

$$\alpha(p) = \frac{\sum (Lpf \times Wf)}{\sum Wf} \tag{7.3}$$

7.4.3 Impact on the Environment, Variable β

As with variable α, variable β corresponds to the existing relationship between the various parts (p) of the product, and is named *Lpp*. That variable can also be achieved on the sub-assembly level. The modification of one part of the system will impact the other parts of the system. Just as we defined earlier, the impact of using a collaborative tool in the development of the design process will be more significant if the modified part has a lot of links with the other parts of the product. The definition of the link between the parts is achieved with the DSM representation.

Consequently, the calculation of the impact on the environment is the ratio of the sum of the links and the sum of the other parts of the system (Formula 7.4).

$$\beta(p) = \frac{\sum Lpp}{\sum p} \tag{7.4}$$

Thus, the two variables α and β represent the significance rate of the parts within the system, and is expressed as a percentage. In order to test our modification rate calculation, the next section outlines a case where this is applied.

7.5 Rating the Modification

7.5.1 Functional Definition of the Part, Variable α

We calculated the modification rate on a case of application. This case was made by the mechanical engineer of the IRTES SeT laboratory, and is detailed in section 7.4.1. In this project, we have stored each version of the CAD model all throughout the design process. A new version of the model was created after every design meeting.

The first step of this calculation is to define the link between the parts and the functional requirement through the DMM. We know that the number of parts change during the design process so this calculation has to be made for each CAD model version (Table 7.1). In this table, f_n represents the functions of the design project requirement and p_n represents the various parts of the CAD model. We put a 1 in the case when there is a link between the part and the function. We can highlight in Table 7.1, that after the design meeting, some parts can be added and the linking of the parts with the functions has changed.

Table 7.1. Relationship between the parts of the CAD model and the functions for version 1 and 2

V1	f1	f2	f3	f4
p1	1			
p2	1		1	
p3		1		1
p4		1		

V2	f1	f2	f3	f4
p1	1			
p2	1		1	
p3		1		1
p4		1		
p5			1	1

The second step of the calculation is the weighting of the functional requirement. This weighting is based on the value engineering method. In this case, there are five criteria with each level of flexibility (see Table 7.2).

Table 7.2. Link between the flexibility and the weight of the criteria

Function	Criterion	Flexibility	Weight
f1	C1	F0	1
f1	C2	F0	1
f2	C3	F2	0.5
f3	C4	F1	0.8
f4	C5	F3	0.2

Thus, with Table 7.2, we notice that the principal function f1 has two value criteria which are both non-flexible (F0), where the three other functions are more flexible. Contrary to the other functions, the first and the third functions are highly weighted. The calculation of the variable α can be calculated for each version (Formula 7.5, Formula 7.6)

$$\alpha_{V1}(p1) = \frac{1 \times 1 + 1 \times 0.5}{1 + 0.8 + 0.5 + 0.2} = 0.6 \qquad (7.5)$$

$$\alpha_{V2}(p1) = \frac{1 \times 1}{1 + 0.8 + 0.5 + 0.2} = 0.4 \qquad (7.6)$$

Thus, in the first CAD model version, part 1 is linked with f1 and f2 and has a significance rate of 60% (7.5), while part 3 is only linked with f4 and has a significance rate of 8%. Consequently, the modification of part 1 will impact the design process more than part 3. In the second version, the modification of the CAD model with the use of a collaborative tool in the design meeting has had an impact on the parts. Indeed, the decisions made during the design meeting have created a new part and have modified three parts of the CAD model. Thus, the significance rate for part 1 has changed from 60% to 40% (7.6), and from 8% to 28% for part 3.

Once the significance rate of the parts of the CAD model is calculated, it is possible to outline its development. Figure 7.2 represents the development of the significance rate for each part of the model in each version.

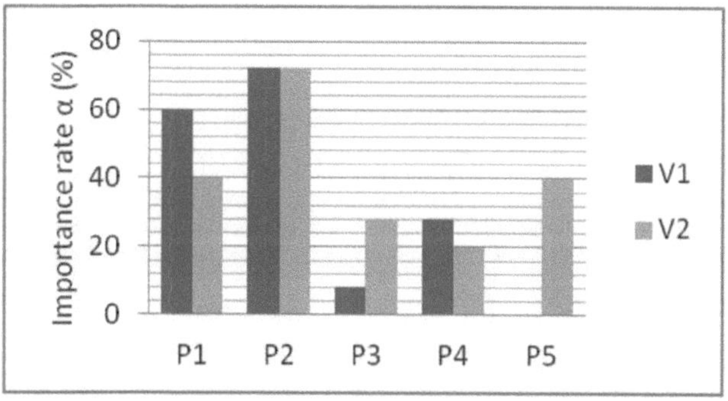

(in graphic above, "Importance rate" should be "Significance rate")

Figure 7.2. Development of the functional significance rate α of the parts during the design project

We can see in Figure 7.2, that the significance rate of the part changes when there are modifications. Thus, part 1 and part 2 are the most significant parts in version 1, while in version 2, part 5 is as significant as part 1. So, the modification carried out during the design meeting has had an impact on a lot of parts and has significantly changed the CAD model. In the case of removing a part, the other parts would be impacted, and we could observe these changes on a graph.

7.5.2 Impact on the Environment, Variable β

Applying the formula for variable β uses the same methodology as the functional definition, variable α. In this case, the DSM of the relationship between the different parts of the CAD model is achieved. This matrix is symmetrical, and the diagonal remains blank (Table 7.3).

Table 7.3. Relationship between the parts of the CAD model for version 1 and version 2

V1	p1	p2	p3	p4
p1		1	1	
p2	1			1
p3	1			
p4			1	

V2	p1	p2	p3	p4	p5
p1		1			1
p2	1				1
p3				1	
p4			1		1
p5	1	1		1	

We observe in Table 7.3, that one part was added and the link between the others parts has changed. In the first version of the CAD model, we can calculate the significance rate of part one over the other parts (Formula 7.7).

$$\beta_{V1}(P1) = \frac{1+1}{3} = 0.66 \tag{7.7}$$

Figure 7.3 represents the calculation of variables α and β and the significance rate of the CAD part.

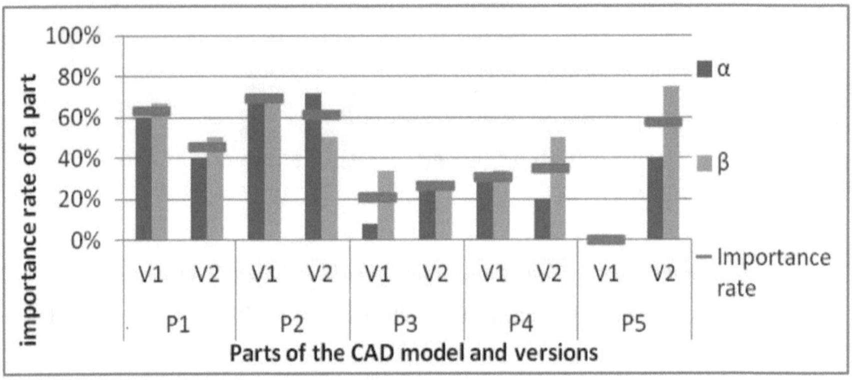

(in graphic above, "Importance rate" should be "Significance rate")

Figure 7.3. Development of the functional significance rate of the parts during the design project

We can see in Figure 7.3, that parts 1 and 2 are the most linked parts in the first version, and then part 5 becomes the most significant part. Indeed, in the second version, part 5 becomes the major part of the system because it is linked with 75% of the total parts. Thus, its modification should impact most of the system parts, however only 40% of the functions concern this part. In the same way, part 4 is linked with half of the parts, but only represents 20% of the overall functions.

Consequently, calculating the significance rate of the CAD parts, firstly gives us information about the the impact that a modification can have on the CAD model's development. Indeed, major modification carried out on mother parts will impact more on the development of the project. We can clearly see that the significance of the CAD parts changes all throughout the design process, and has to be studied in order to understand the development of the data becoming complete.

This paper only focuses on calculating the significance rate of the part, which is only one criterion to define the modification rate of a CAD model. Other criteria of the modification rate, are the significance of the modification itself and the modification type. So, the next step of this approach is to calculate the modification rate based on the significance rate, defined in this paper, over the type of modification and over the significance of the modification. Consequently, knowing the modification rate of a CAD model after using a collaborative tool during a design meeting, makes it possible to understand the impact of the chosen collaborative tool. In order to be more effective, the calculation has to be made automatically using a PLM system and demonstrated directly to all the stakeholders of the project. However, this approach is limited, as modifications made on the CAD part can be caused by other factors other than the collaborative tool used, for instance the expertise of the stakeholder or the development of the design requirement. So, the evaluation of the use of collaborative tools is a multi-criteria approach and other factors must be considered.

7.7 Conclusion and Outlook

This paper has demonstrated an evaluation approach based on the CAD part significance rate definition, all throughout the design project. This evaluation approach is a first step of a more "overall" evaluation of collaborative tools during the design process, based on the modification rate of each part of the design project. In the prior art, we have seen that there are numerous collaborative tools used all throughout the design process in various forms, depending on the users' need. Their use allows decisions to be made which impact the project through product data modifications. The main current problem is that the evaluation of these tools is principally based on their usability and their results in a given time. Our hypothesis is that the use of collaborative tools impacts each piece of the project's data, and their evaluation can be done in accordance with this criterion. Consequently, this approach is additional to the current evaluation method. Thus, in order to evaluate these collaborative tools, we propose to observe their impact on the three layers of the project: the social, process and product layers. This article focuses on the product layer corresponding to the project's data. In this layer, the use of collaborative tools is only observed through the project data. Thus, we have proposed two variables which define the significance rate of a part as regards the relationship among the parts and between the parts and the requirements.

The next step of this research is to mathematically define the modification rate based on the significance rate of the part and the type of modification. Also, the expertise of the various actors in defining the requirement list and in its weighting has to be included in this composition. The weight of the various criteria can also develop throughout the whole design process in accordance with the development of the requirement list. Moreover, this research must consider the three layers of the design project: the social, process and product layer in order to have a complete evaluation of one collaborative tool.

Acknowledgement

This research has benefited from the help of Jérôme Maysse, a mechanical engineer at the IRTES-SeT laboratory of the University of Technology of Belfort-Montbéliard.

7.9 References

Al-Khatib, A., Mahdjoub, M., Bluntzer, J. B., Sagot, J. C. (2013) A Tool Proposition to Support Multidisciplinary Convergence in Immersive Virtual Environment: Virtusketches. In Smart Product Engineering, pp. 795-804. Springer Berlin Heidelberg

Antunes, P., Herskovic, V., Ochoa, S. F., Pino, J. A. (2012) Structuring dimensions for collaborative systems evaluation. *ACM Computing Surveys (CSUR)*, *44*(2), p. 8

Bennes, L., Bazzaro, F., Sagot, J. C. (2012) Virtual reality as a support tool for ergonomic-style convergence: multidisciplinary interaction design methodology and case study. In Proceedings of the 2012 Virtual Reality International Conference, p. 24. ACM

Bluntzer, J. B., Sagot, J. C., Mahdjoub, M. (2011) Knowledge Based Engineering Approach through CAD Systems: Results of a 2 Years Experimentation in an Industrial Design Office. In Global Product Development, pp. 545-551. Springer Berlin Heidelberg

Clarkson, P. J., Simons, C., Eckert, C. (2004) Predicting change propagation in complex design. *Journal of Mechanical Design*, 126(5), pp. 788-797

Couto, V., Mani, M., Lewin, A. Y., Peeters, C. (2006) The globalization of white-collar work. Duke University and Booz Allen Hamilton (Durham, NC: Duke CIBER)

Cross, N. (2000) Engineering design methods: strategies for product design (Vol. 58). Chichester: Wiley

Defays, A., Safin, S., Darses, F., Mayeur, A., Rajeb, S. B., Lecourtois, C., Guéna, F., Leclercq, P. (2012) Invisible computer for collaborative design: evaluation of a multimodal sketch-based environment. *Work: A Journal of Prevention, Assessment and Rehabilitation, 41*, pp. 3494-3503

Dorta, T., Kalay, Y., Lesage, A., Pérez, E. (2011) Design conversations in the interconnected HIS. International Journal of Design Sciences and Technology, 18(2), pp. 65-80

EN N. 1325-1 (1996) Vocabulaire du management de la valeur, de l'analyse de la valeur et de l'analyse fonctionnelle-Partie

French, M. (1998) Conceptual design for engineers. Springer

Geyer, F., Pfeil, U., Budzinski, J., Höchtl, A., Reiterer, H. (2011) Affinitytable-a hybrid surface for supporting affinity diagramming, pp. 477-484. Springer Berlin Heidelberg

Giffin, M., de Weck, O., Bounova, G., Keller, R., Eckert, C., Clarkson, P. J. (2009) Change propagation analysis in complex technical systems. Journal of Mechanical Design, 131

Hartmann, B., Morris, M. R., Benko, H., Wilson, A. D. (2010) Pictionaire: supporting collaborative design work by integrating physical and digital artifacts. In Proceedings of the 2010 ACM conference on Computer supported cooperative work, pp. 421-424. ACM

Jarratt, T. A. W., Eckert, C. M., Caldwell, N. H. M., Clarkson, P. J. (2011) Engineering change: an overview and perspective on the literature. Research in engineering design, 22(2), pp. 103-124

Ostrosi, E., Haxhiaj, L., Fukuda, S. (2012) Fuzzy modelling of consensus during design conflict resolution. Research in Engineering Design, 23(1), pp. 53-70

Pasqual, M. C., de Weck, O. L. (2012) Multilayer network model for analysis and management of change propagation. Research in Engineering Design,23(4), pp. 305-328

Pahl, G., Beitz, W. (1996) Engineering design: A systematic approach, K. Wallace, Ed., Springer

Pugh, S. (1991) Total design: integrated methods for successful product engineering Wokingham: Addison-Wesley

Saint-Marc, L., Callot, M., Reyterou, C., Moly, M., Girard, P., Deschamps, J. C. (2004) Toward a data maturity evaluation in collaborative design processes. In Proceedings of the 8th International Design Conference DESIGN 2004 pp. 69-76

Stark, J. (2011) Product lifecycle management (pp. 1-16). Springer London

Steward, D. V. (1981) The design structure system: a method for managing the design of complex systems. Engineering Management, IEEE Transactions on, (3), pp. 71-74

Ulrich, K. T., Eppinger SD (1995) Product design and development (Vol. 384). New York: McGraw-Hill

Van Dijk, J., Vos, G. W. (2011) Traces in creative spaces. In Proceedings of the 8th ACM conference on Creativity and cognition, pp. 91-94. ACM

Wodehouse, A., Ion, W. (2010) Digital information support for concept design. CoDesign, 6(1), pp. 3-23

Chapter 8

Scrum in the Traditional Development Organization: Adapting to the Legacy

N. Ovesen and A. F. Sommer

8.1 Introduction

During the last couple of years, the application of Scrum as a project management framework has been broadened from initially belonging to the software domain. Now companies within the field of traditional product development are starting to implement Scrum in an attempt to improve their development efforts with respect to resource efficiency and speed. But as a process control model, Scrum is radically different from those traditional process control models, which are typically favoured among the stakeholders in the companies' management levels. Traditional and highly defined models, such as Stage-Gate, form a solid backbone in the development organizations of thousands of companies as they offer a formal and long-term project plan with critical reviews, logical and sequential phases, and a transparent distribution of responsibility (Browning and Ramasesh, 2007).

However, the famous and almost 30 years old quote from Takeuchi & Nonaka (1986) about the essential need for speed and flexibility in today's fast-paced and fiercely competitive world of commercial new product development, was just one of the first of many to question the strict plan-your-work-work-your-plan strategy. Now, in some companies within the domain of integrated product development, Scrum is seemingly being integrated as an addition to these established process models rather than as a substitute.

This paper is based on a case study of a series of Danish companies working with Scrum as a project management framework within a traditional Stage-Gate setting. All companies have been working with the implementation of Scrum for a period from one to four years, and all the companies are developing products consisting of software, firmware, hardware, mechanical designs, and several other domains.

The main answer sought answered in the paper is this: *What are the main organizational challenges of implementing Scrum in Integrated Product Develop-*

M. Schabacker et al. (eds.), *Modelling and Management of Engineering Processes*, DOI 10.1007/978-3-662-44009-4_8

ment environments where formal process models already exist? As an answer to the question, the paper presents a series of challenges identified through interviews and video observations from seven companies. The paper focuses on challenges experienced in relation to the organization and the management. Numerous challenges exist in relation to the actual development effort; these are *not* covered in this paper.

The rest of this paper is structured as follows: Section 8.2, *Theoretical Overview*, covers the basic structure of the Scrum framework. Section 8.3, *Methods*, gives an overview of the cases and the methods used to obtain data from them. Section 8.4, *Results*, presents the most important data from the cases, and finally Section 8.5, *Analysis and Discussion*, lays out the principles, relationships and generalizations that can be derived from the results.

8.2 Theoretical Background

This section covers the concepts of Agile Development and Scrum. It furthermore gives a brief overview of the Stage-Gate process model as well as Integrated Product Development.

8.2.1 Agile Development and Scrum

Agile Development, as a term, was coined early in 2001 during a two-day meeting between seventeen people gathering at Snowbird Ski Resort in the Wasatch Mountains of Utah (Highsmith, 2013) The gathered people were representatives from various surfacing disciplines in software development trying to establish a common ground and explicate a united stance in the worldwide software development community. The outcome of the summit in this extraordinary place was The Manifesto for Agile Software Development. Table 8.1 below shows the value set from The Agile Manifesto of Software Development. The four statements clearly make up with the command-and-control development processes in traditional development (Suscheck and Ford, 2008).

Table 8.1. The value set of Agile Development (Highsmith, 2013)

We are uncovering better ways of developing software by doing it and helping others do it. Through this work we have come to value:

- **Individuals and interactions** over processes and tools
- **Working software** over comprehensive documentation
- **Customer collaboration** over contract negotiation
- **Responding to change** over following a plan

That is, while there is value in the items on the right, we value the items on the left more.

Scrum is regarded as one of the process models adhering to the manifesto above. Takeuchi & Nonaka first mentioned the concept of Scrum in relation to development as early as in 1986. The term originates from the strategy used in Rugby for getting an out-of-play ball back into play. The name was chosen because of the similarities between this game and product development – both are adaptive, quick, self-organising, and have few rests (Schwaber and Beedle, 2002). With scrum the emphasis is on an empirical process rather than on a defined process (Larman, 2004).

Rather than ultimately deciding variables such as requirements, resources, technologies, and tools only at the beginning of a project, the development phase is organised in short iterative cycles called Sprints, where these variables are continuously revised and thoroughly controlled (Figure 8.1). A Sprint focuses on the development of only a few collectively chosen features in the product backlog list. Scrum emphasises self-organising teams and most importantly frequent Scrum-meetings between all the team members. Each Sprint ends with a Sprint review and a revision of the backlog, and the development phase ends when the requirements are completed through several Sprints cycles (Schwaber, 2009).

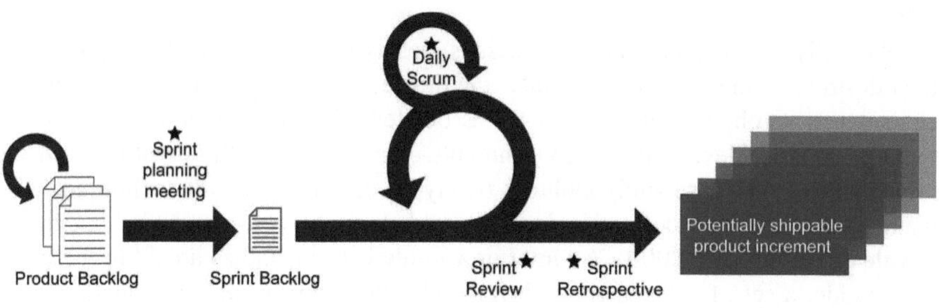

Figure 8.1. The Scrum framework with the big circular arrow representing the 2-4 week Sprint

8.2.2 Stage-Gate and Integrated Product Development

Scrum is often seen as a contrast to process models such as the original Stage-Gate model by Cooper (1979). Stage-Gate offers a linear structure with stages and gates that together forms the process from idea to launch of a product. Essentially the gates act as the quality control checkpoints, where fundamental questions are asked about the state of the project (Cooper, 2011). In order to pass a gate all defined tasks must be completed (Browning and Ramasesh, 2007). All companies in the case study presented in this paper make use of various versions of Stage-Gate models as the main process control model.

As mentioned earlier, all companies furthermore combine several types of professional practices in their respective development projects. This is done in an integrated manner with overlapping and parallel workflows, also known as Integrated Product Development (IPD). According to Gerwin and Barrowman (2002), IPD contrasts sequential development and regards interactions between different functional disciplines highly important.

8.3 Methods

The present study is primarily based on the case study method (Yin, 2003). According to Yin "a case study is an empirical inquiry that a) investigates a contemporary phenomenon within its real-life context, especially when b) the boundaries between phenomenon and context are not clearly evident. (…) In other words, you would use the case study method because you deliberately wanted to cover contextual conditions" Yin (2003).

The main criterion as to the choice of companies is the presence of experience with the Scrum framework. The companies are primarily identified through the "Scrum Denmark forum" at LinkedIn.com, and it is ensured that each of the case companies are developing products that requires a broad range of disciplines and that their respective development environments are conducting Scrum to a certain extent.

Naturally, the data collection process is closely depending on which type of data is desired. In the case of this study, the desired data are primarily of qualitative character as its object is the challenges associated with implementing Scrum in integrated product development environments *experienced* by the employees in the organizations. The case study includes two types of data collection techniques: The main part of the data collection is conducted through *semi-structured interviews* (Kvale & Brinkmann, 2009). An interview guide is used, and as an aid to the interview guide, a set of A3-posters is developed, illustrating visual representations of some of the questions. Interviewees are selected based on two criteria:

1. Experience with participation in projects conducted through Scrum
2. Interviewees should represent different Scrum roles

Video observation is used as a supplementary technique for collecting data in some of the seven cases.

8.3.1 Case Overview

In Figure 8.2 a basic overview of the seven cases is given. It includes information about general scrum experience, disciplines, organization and process models of the respective case company.

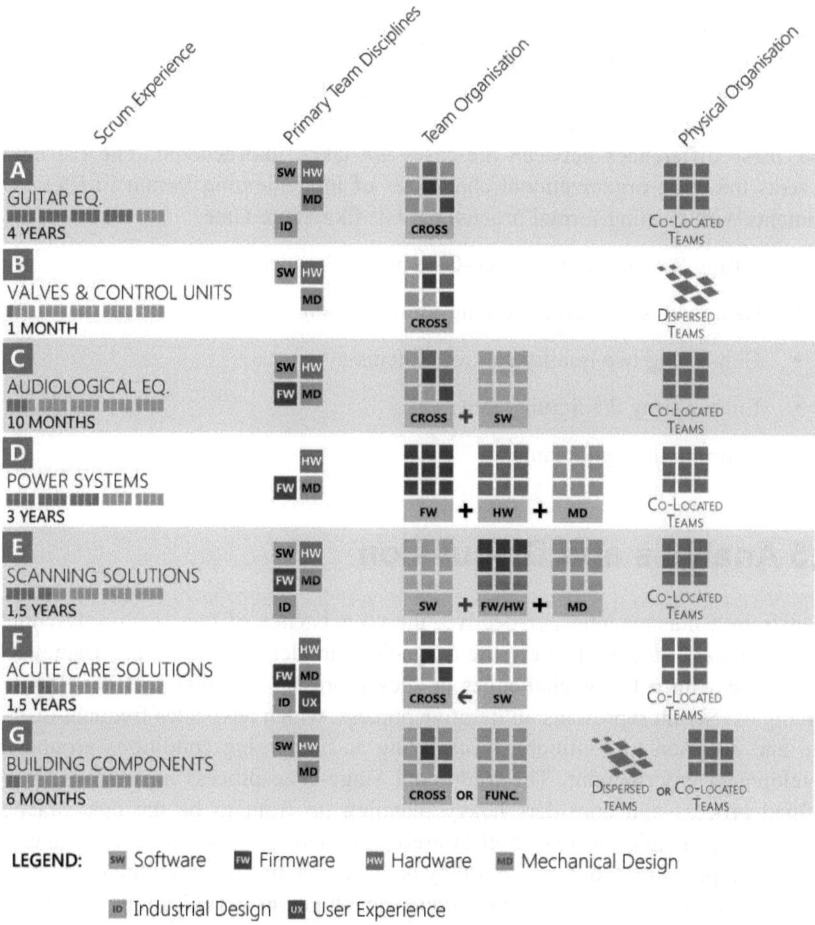

Figure 8.2. Overview of the cases included in the study

All interviews have been carried out at the respective companies over a period of four months, and all interviews have been audio-documented and partly transcribed for later analysis.

The analysis of the data has been done in two steps. The first step focused on establishing a comparative overview of the cases and on converging the challenges identified through the case descriptions into a set of themes. The second part of the analysis focused on the identified themes through a structured scheme.

8.4 Results

Each case has been investigated and analyzed with respect to its compliance with the Scrum framework. Not all companies have implemented all parts of Scrum, and each of them have chosen to integrate roles, implement Scrum events, and use Scrum artifacts in their own way. But it is not only the level of compliance that differs from the seven companies. Also the organizational setups in which Scrum is implemented form a rather diverse picture. In the following analysis and discussion, these differences between the cases are taken into account. The list below presents the main organizational challenges of implementing Scrum in IPD environments with existing formal process models like Stage-Gate.

- Fitting Scrum into the Stage-Gate process model
- Balancing short-term and long-term planning
- Combining two quality control systems
- Empowering the Scrum team
- Forming the right teams

8.5 Analysis and Discussion

Seen from a management perspective, the co-existence of two seemingly contradicting process models, namely the Stage-Gate model and the Scrum framework, does come with a lot of challenges as they represent two different development paradigms. Scrum represents an iterative process, which advocates frequent inspection and adaption to continuously changing and emerging conditions around the development environment. The traditional Stage-Gate process advocates a rather defined process and considers heavy planning up front to be the best practice. Clearly, in principle the two models are contradicting – also in several other respects than planning schemes – but they *do* exist side by side in all the seven cases, and this indicates that integrated product development organizations can benefit from Scrum *in practice*.

The study of the seven companies shows that more than 30 different types of challenges exist. These challenges include both issues related to the practical development from a design and engineering perspective, and issues related to the set-up and structure of the organization. As mentioned earlier, this paper specifically focuses on the management challenges related to implementing Scrum in the existing Stage-Gate-oriented matrix organization, and within this framing the main challenges are analyzed and discussed in the following sub-sections

8.5.1 Fitting Scrum into the Stage-Gate Process Model

When trying to implement Scrum in a traditional organization, it is seemingly a combination of two contradicting planning paradigms. The best practice when planning a development project in a traditional organization has typically been to plan in detail up front; but Scrum proposes a significantly shorter and continuously moving horizon for detailed planning. This has often been characterized as two conflicting extremes: Scrum versus Stage-Gate or Waterfall – or empirical versus defined process control, as shown by Schwaber (2009). However, in all seven cases Scrum is a supplement to an existing process model in various versions of Stage-Gate. This seemingly works relatively well as none of the cases indicates something else. In a couple of cases, it has been mentioned that Scrum is conducted "below the radar" or without any relation to the Stage-Gate above it. In other cases, as for instance Case G, Scrum is intentionally used as the lowermost process management framework, which is fitted into the sub-sections of the "mid-level" VVSM process. This relation between the Stage-Gate, the VVSM and the Scrum cycles is seen in Figure 8.3.

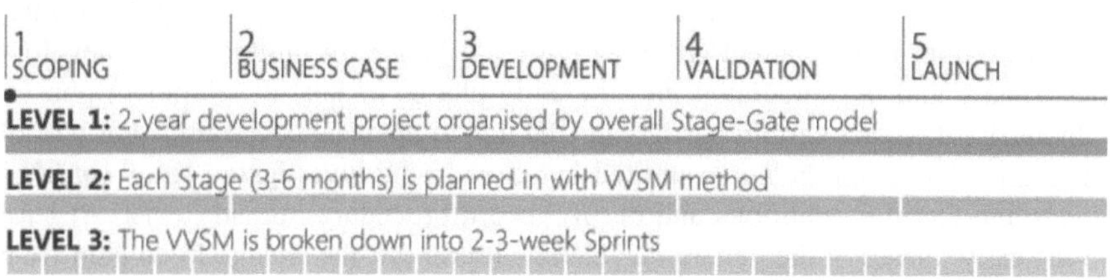

Figure 8.3. The ideal relationship between Stage-Gate, VVSM and Scrum in Case G

When Scrum is integrated into a development process controlled through an existing Stage-Gate model, the paradigmatic divide may not always be a big challenge in practice. However, in the investigated cases, the Scrum framework rarely extends to other stages than the development stage, which is shown in Figure 8.4 below.

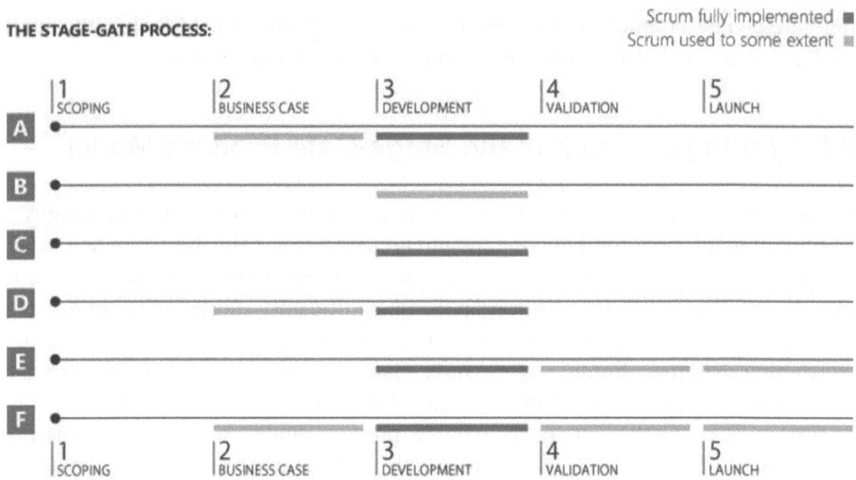

Figure 8.4. The cases from A to F, showing Scrum activity in mainly the development phase. Case G is different as shown in Figure 8.3.

The general picture in Figure 8.4 is contrasting one of the main ideas in Scrum. In its original domain, Scrum generally seems to include the whole development process from the early and broadly defined concepts to the launch and maintenance of the software or system. This case study shows that Scrum, when applied to Integrated Product Development (IPD) activities, is generally limited to the development phase and *not* integrated into the early phases or post-development phases. Obviously, there are certain barriers in IPD due to physical constraints, but part of the reason for Scrum being so limited in Scope in the investigated cases *could* be due to the fact that traditions and deeply rooted procedures already exist. At least in the early phases of the general development process, the change in domain from software to IPD does not seem to ad any domain-specific limitations to the Scrum framework.

8.5.2 Balancing Short-term and Long-term Planning

The main artifact in Scrum is the Product Backlog – a document equivalent to the typical requirements specification, managed by the "Product Owner" and continuously groomed throughout the development process. The Product Backlog is, however, different from the requirement specification in several aspects, and the grooming of it includes the process of revising it according to the present situation and new insights. The Product Backlog is, in contrast to a traditional specification, continuously re-prioritized and detailed in the same pace as the development moves forward.

In Case A the Product Backlog is implemented and managed by the traditional project manager, and it has not been a significant problem to integrate the two process models on an overall basis. However, the interviewees argue that Product

Backlog from Scrum requires another mindset in regards to the practical planning of the process:

> *"We need to have an idea about what is going to happen six months into the future. Only an idea – we must not by any means go into detail on it, but we need to be able to communicate it in order for people to make the right decisions today."*
>
> Interviewee 1, case A

In most of the investigated cases, the Product backlog is not revised as prescribed by the Scrum Guide. Instead, the general product requirement specifications of the respective development projects are used as the point of departure for the work packages specified to the separate Sprint cycles. But whereas this seems to work well in most cases, the continuous revision of the Product Backlog, which allows for the inclusion of new insights and re-prioritizing of existing requirements, is typically left out.

8.5.3 Combining Two Quality Control Systems

As Blessing (1993) argues, the quality of the product strongly depends of the quality of the process, and the Stage-Gate process provides a clearly defined process, which according to Phillips *et al.* (1999) is an important part of this. Furthermore, quality management systems such as the ISO 9000 family of standards fit well with the rigid and plan-driven process model of Stage-Gate (McMichael and Lombardi, 2007).

Scrum, as a contrast, has been criticized for its lack of critical design reviews like the gates in the Stage-Gate process (Boehm and Turner, 2005). Scrum promotes a significantly shorter development cycle that ends with a Sprint Review, counting as the quality control in Scrum. In the cases of this study, the Sprint Reviews do not substitute the quality control of the Stage-Gate model. It rather supplements it as an internal and frequent quality check of the development effort within the Scrum team, and is not considered as a critical quality review by the management level stakeholders.

Reducing the Sprint Review to an internal quality check of the delivered product increment in the develop team may not be a challenge in itself as long as requirements are met and not changed, but when corrective actions and changes to the product requirements are needed, the team may not be able to act on it as otherwise promoted by the Scrum framework. This is due to its lack of empowerment within the Scrum team and relates to conflicting distributions of responsibility and empowerment between a traditional project management setup and the Scrum framework.

8.5.4 Empowering the Scrum Team

When implementing Scrum in an organization with a traditional management structure it will require some adjustment of one or both solutions. Just as traditional management, Scrum has a portfolio of formal roles that are required in order to be conducted properly. Scrum roles and traditional management roles, such as the project manager role, have a certain overlap. While the responsibility for the execution of projects is born by the project manager in the traditional management model, project management is a shared responsibility in the self-organizing Scrum development team. The Scrum Master is only facilitating the process, and the Product Owner represents the customer and the business perspective.

In most of the investigated cases, the Product Owner is absent, and often the traditional management model and the Scrum framework only flank each other in the presence of one single person playing the dual role as both Project Manager and Scrum Master. However, according to the data obtained through the cases, this dual role does not seem to cause any significant problems. On the other hand the role of the Product Owner – perhaps more rightly the absence of it – seems to cause some frustrations in the development teams in at least two of the cases:

> *"There have been a lot of battles and they have taken a huge amount of time – at that point it would have been nice if an actual Product Owner would have taken those decisions."*
>
> Interviewee 1, Case D

> *"As a Scrum Master I miss some inputs to a Product Backlog. In reality our projects operate without Product Backlogs."*
>
> Interviewee 1, Case F

In both cases the lack of a Product Owner results in frustrated Scrum Masters and Development teams, and without a Product Owner taking the responsibility of managing the Product Backlog, this task trickles down to the development team. Due to the long lines of command in the surrounding organization, the teams in Case D found it difficult to maneuver and take the necessary decisions in the extensive hierarchical organization:

> *"Our Product Line Manager, who owns the product when it goes to market and who talks to marketing and customers, needs to be located close to the Scrum team in order to have the essential and daily communication [with the team] while managing this Product Backlog. Right now he sits in India and is extremely difficult to reach."*
>
> Interviewee 2, Case

The Scrum framework does not give any guidance to how it should be implemented in a large organization. This is decided by the management, and to some extent also the development environment in which it is implemented. The frustration found in the quotes above is not present in the case A and C, as both of them are deploying Product Owners with the responsibility of grooming the Product backlogs.

8.5.4 Forming the Right Teams

The official Scrum guide promotes cross-functional teams, and in software development, which is the original domain of the Scrum guide, cross-functionality means a mix of disciplines *within* the software domain. In integrated product development cross-functional teams entail a significantly larger variety of involved disciplines. This fundamental difference in the transition from software development to integrated product development clearly has some consequences to the teams in the investigated cases. Extremely cross-functional teams have certain communication issues, which, in some cases, lead to a drop in motivation. The fundamental difference between the two domains is that in software, team members have software development as a unifying discipline; the majority of a cross-functional software development team has to some extent overlapping competences. This is not necessarily the case in cross-functional teams in integrated product development, which may very well include software, firmware, hardware, mechanics, industrial design and more.

> *"In the software silo there is a critical mass of developers with the same competences. In our project team we can easily be just one electro technician, one chemical engineer and one mechanical engineer. They just really can't share tasks other than getting the coffee."*
>
> Interviewee 1, Case F

As indicated in the quote above, the cross-functional team is not able to achieve the same collaboration synthesis as is achievable in the software development silo. This might very well be the reason why the organization is slowly starting to build up competence silos of mechanics, firmware and chemistry similar to the software development silo. In Case D, which describes three synchronized and functionally divided teams, interviewee 2 argues that even with functional teams the close collaboration may be difficult, due to the need of a broad variety of experts:

> *"I have six developers on the Mechanics team, all with very different competencies, so this is not true Scrum. I cannot just put anyone onto a certain task, and that's a challenge. One person does all the plastics and another does this and that."*
>
> Interviewee 2, Case D

Together, the two quotes above reflect some of the difficulties in composing teams in integrated product development. It can be argued that two different paths can be taken, when composing Scrum development teams for integrated product development of a certain complexity:

- Several parallel functional teams
- Large and extremely cross-functional teams

In this regard, the challenge is to balance cross-functionality with a certain critical mass of homogeneous competence in the development teams.

In both Case D and Case F multiple Scrum development teams are working together on the same project. However, the two companies are handling this with varying success. While the teams in Case D are systematically communicating through both Scrum Masters and System architects, Case F has not yet established a formal way of synchronizing collaborating teams.

8.6 Conclusion

This paper presents a series of challenges related to the implementations of Scrum in IPD environments where existing stage-gate process models exist. The challenges are identified in a case study of seven Danish companies working with product development in combination with Scrum.

The case study shows that Scrum in general is used as a tact-enabling tool for carrying out development tasks in an efficient manner within the existing Stage-Gate-controlled development process. In the investigated cases, Scrum is an additional framework that is only utilized to some extent. However, it is despite the focus on challenges presented in this paper, clear that the development environments are benefitting from the Scrum framework in many ways and it is argued by all interviewees that Scrum is a positive and highly beneficial addition to their development process. With this paper, we want to initiate a discussion about how the Scrum framework can be even more beneficial to the development environments by focusing on the challenges that still need to be addressed.

8.7 References

Blessing, L. T. M. (1993) A Process-based approach to computer supported engineering design. Proceedings of the International Conference on Engineering Design '93: 1393-1400

Boehm, B. W., Turner, R. (2005) Management Challenges to implementing Agile Processes in Traditional Development Organizations. IEEE Software, 22(5): 30-39

Browning, T. R., Ramasesh, R. V. (2007) A Survey of Activity Network-Based Process Models for Managing Product Development Projects. Production and Operations Management, 16:217-240

Cooper, R. G. (1979) The dimensions of industrial new product success and failure. The Journal of Marketing, 43(3): 93-103

Cooper, R. G. (2011) Winning at New Products – Creating Value through Innovation, 4th edn. Basic Books, New York

Gerwin, D., Barrowman, N. J. (2002) An evaluation of research on integrated product development. Management Science, 48:938-953

Highsmith, J. (2013) History: The Agile Manifesto, http://agilemanifesto.org/history.html, 17-10-2013

Kvale, S., Brinkmann, S. (2009) InterView – Introduktion til et håndværk, 2nd edn. Hans Reitzels Forlag, Copenhagen

Larman, C. (2004) Agile and Iterative Development: A Manager's Guide, Addison-Wesley Professional

McMichael, B., Lombardi, M. (2007) ISO 9000 and Agile Development. Proceedings of AGILE 2007, pp. 262-265

Phillips, R., Neaily, K., Broughton, T. A. (1999) A comparative study of six stage-gate approaches to product development. Integrated Manufacturing Systems, 10(5) pp. 289-297

Schwaber, K (2009) Agile Project Management with Scrum. Microsoft Press, Redmond, Washington

Schwaber, K., Beedle, M. (2002) Agile Software Development with Scrum, Prentice Hall, New Jersey

Suscheck, C. A., Ford, R. (2008) Jazz Improvisation as a Learning Metaphor for the Scrum Software Development Methodology. Software Process Improvement and Practice 13: 439 - 450

Takeuchi, H., Nonaka, I. (1986) The new new product development game, Harvard Business Review, 64(1):137-146

Yin, R. K. (2003) Case Study Research – Design and methods, 3rd edn, 2003. Sage Publications, California

Chapter 9

Business Coaching and Consulting – the Systemic Constellation Approach in Business

C. Burchardt

9.1. Introduction

Business coaching and consulting as a systemic constellation approach is a solution-focused process, which helps leaders of organizations to identify the complex, often informal, relationships and inter-dependencies within their organization and to develop a deeper understanding about the underlying dynamics in a very time- and cost-efficient manner. The systemic constellations are a technique for gaining insight regarding the core of a problem or an issue in a short amount of time. One of the characteristics of a problem is that it fixates out attention and makes it impossible for the people to divert it. Systemic constellations render the inner image of the structure of a problem visible and tangible; through focusing on steps and changes new resources, solutions and opportunities for action are possible.

The effectiveness of this method has become a widely accepted discipline in business consultancy, organizational development and change management. The process serves as a platform to solve business issues and to create future change – e.g. developing a sustainable corporate culture, optimizing the organization's performance and enabling the strengths of individual team members as well as the team as a whole to fulfil their tasks in the most productive way. This paper describes how the systemic constellation approach can be transferred and applied to the field of business.

M. Schabacker et al. (eds.), *Modelling and Management of Engineering Processes*, DOI 10.1007/978-3-662-44009-4_9

9.2. Business Coaching and Consulting

Business coaching and consulting generate different benefits: Fresh perspectives on personal challenges, enhanced decision-making skills, greater interpersonal effectiveness, and increased confidence (Migge, 2011). Different companies are realizing the effectiveness of business coaching in achieving their goals to increase employee satisfaction, improve output and strengthen their bottom lines. As an example business coaching is used to shift a corporate culture to increase productivity by changing it from command and control to collaboration and creativity (Clutterbuck, 2009). It helps to close the gap between generations by increasing engagement and encouraging progress that benefits all parties involved Leadership is strengthened; communication is enhanced; listening is fine-tuned; and the overall organization becomes more effective. Business coaching contains different styles, e.g. systemic constallation approach, hynosystemic coaching, neuro linguistik programmation and other methods. A common procedure of all these styles (Figure 9.1) will be, that the coach support the coachée to change the mindset.

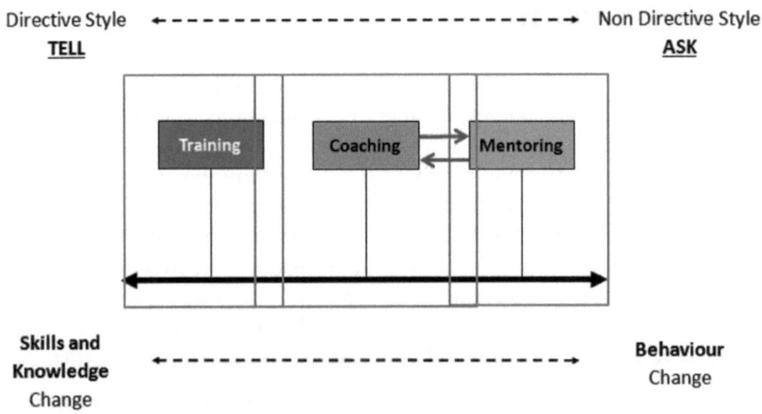

Figure 9.1. Business Coaching as a helping continuum

9.3 Systemic Constellation Approach

Often, organizations and companies face complex problems:

- the company does not have an adequate development,
- the clients disappear,
- it is not possible to innovate the products,
- the employees are in permanent conflict,

- the merger of two companies does not allow the business to develop,
- the employees leave without any real reason, etc.

A rational analysis visualizes trails or indicates part of the problem, but never an overall view. Through a skilful use of systemic constellations, it is possible to discover hidden dynamics that are in operation in the organizations and companies.

9.3.1 Feature of Systemic Constellation - Phenomenological Constellation

The defining feature of systemic constellation is the phenomenological constellation which was originally developed for families and then applied to other systems as systemic constellation work such as businesses, organizations, education and the medical field (Hellinger *et al.*, 1998). The systemic constellation approach combines systemic understanding [systemic theory acknowledges that all elements within a system, such as group members, are interdependent and interactive], using of representatives work [using representatives to represent family members and or elements of larger systems and even concepts] and the phenomenological method [loosely interpreted as being a technique of acknowledging what is without preconception or prejudice]. The combination of these three streams allows unique interweaving of the process in many fields including environmentalism, quantum physics, psychology and spirituality.

Basics idea behind the Phenomenology

- **The whole is greater than the sum of its parts.** So a living human body is more than just its organs and limbs. It depends on how they connect and interact. Similarly a family or organization is more than just a number of individual members put together. It also depends on how they link and interact.
- **All elements in a system are interdependent.** Changes in one element result in changes in all the others. In organic systems such as the human body the system works to maintain equilibrium. This process is called homeostasis. In this situation survival of the system takes priority over the survival of the component parts. So when a person is exposed to extreme cold the vital organs will be protected while the extremities may be permanently damaged by frostbite. Similarly in families an individual may be sacrificed for the sake of the system

9.3.2 Systemic Constellation - General

The Systemic Constellation approach (Sheldrake, 2003) began in the field of psychotherapy but in the course of the last several years it has reached far beyond this scope and is currently on its way to making significant contributions to the body of social work, therapy, and many other medical fields such as homeopathy, as well as offering effective solutions for strengthening businesses and organizations in several areas such as team building and development, and other consulting tasks (Sparrer, 2011). The approach was further developed by Insa Sparrer (Sparrer, 2009), Matthias Varga von Kibéd (Kibéd, 2006) with influence of Bert Hellinger and Gunther Weber (Hellinger *et al.*, 1998). Later on business issues were adressed of Gunthard Weber (Weber, 2004) and Varga von Kibéd.

In general the systemic constellation approach understand organizations and companies as a complex, living and learning network of relations which are interwoven with each other in many ways. This may be through contracts and official and unofficial hierarchies but also simply through the relationships the employee of a company or organization have with each other. Complex living systems behave in many ways like a mobile (Figure 9.2).

Figure 9.2. Mobile principle of object movement – all actions influences others

Wikipedia specify a mobile as a type of kinetic sculpture constructed to take advantage of the principle of equilibrium. It consists of a number of rods, from which weighted objects or further rods hang. The objects hanging from the rods balance each other, so that the rods remain more or less horizontal. Each rod hangs from only one string, which gives it freedom to rotate about the string. If one of the mobile objects get movement all other balanced objected get movements in sometimes unpredictable ways.

This is one of the reasons why the systemic approach, instead of focusing just on the elements themselves, emphasizes the relationships between elements. By zooming out a better picture of the whole are visible so diagnosis becomes more precise and takes the whole into account. When the gaps are identified, it is possible to fill them in a balanced way for the whole. Adding to this, the process has to utilize to bring out the best in business. By creating possible contexts in which win-win situations are more likely to happen it also invites long-term success in business.

9.3.3 Systems Constellation - Basic Principle

The Systemic Constellation Basic Principle are:

- Respect is the most important principle (what is, must be allowed to be).

- Everyone in the system has a right to her or his place (right to belong).
- There must be a balance of giving and taking between individuals, between individuals and the system and between different parts of the system.
- The system requires that certain priorities and orders of precedence should be observed. These include length of service, specialist skills, qualification, functional hierarchy, competence and particular stakes in the system.

9.3.4 Systemic Constellation - Formats

The usage of systemic constellation work depends of the environment situation. Different formats/types of constellations are developed. They are used according the cases, the issues and the sensibility of the facilitator. Main types of constellations formats are:

- **Normal** (initially developed by Bert Hellinger).
- **Blind or hidden** (in which there is no information about the issue or theme).
- **Structured** (developed by Matthias Varga von Kibéd, that includes some of the followings and they are predefined formats of Constellations for specific type of questions).
- **Problem** (can be done hidden, with representatives for the resources, the obstacles, the solution/objective and focus (client); it is also possible to include a representative for a medium term objective and for the hidden interests).
- **Diagnostic** (useful when there is not a specific question or to see the dynamics or to get an organization diagnosis).
- **Project** (for project analysis, quantifying feelings and sensations).
- **Tetralema** (very useful for dilemmas or when it is difficult to decide between two alternatives or solutions).
- **Value Triangle** (separating the individual main values and making an independent analysis for each one, in order to get to its integration).
- **Butterfly** (project study and its evolution).

9.3.5 Systemic Constellation - Paradigm

The systemic paradigm abandons the mechanistic model of the world and the belief in objectivity, applies the principle of multiple lenses through which we perceive reality and emphasizes self-monitoring. Examples of the procedure of the mechanistic and systemic view are listed in Table 9.1 (Bodirsky, 2013).

Table 9.1. Differences between mechanistic and systemic world-view

Mechanistic View	**Systemic View**
Objectivity, one truth, unchangeable laws	Construction of reality, many „truths“, thesis
Right-wrong, guilty - not guilty	Context-dependency, utility, connectivity
External control	Self-monitoring, self-organisation
Linear causal chain	Multiple interdependency, feedback loops
Fixed difference, measurable	To be different, change
Linear progress	Development, change and preservation, unblocking, discontinual change
Formal logic, free from contradiction	Integration of contradiction, including
Hard facts, rational relationships	Integration of hard and soft facts (emotions, intuition, processes of communication etc.)
Roles: leader, manager	Roles: impulse generator, enabler, development worker, gardener, coach
Methods: instructive, directive, command, learning through trial and error	Methods: listening, questions, dialogue, discussion, reflection, learning to learn

9.4 The Systemic Constellations Approach in Business Context

All constellations relate to organizations, even a family or a human body is an organization. A system included a number of elements that are connected to one another in a continuously changing relationship. A system can be e.g. any group of people who regularly work, learn, or play together. Within business context, the organization as a whole represents a system. But in companies usually different departments and divisions exist, which also represent systems within the whole - sub-systems that belong to the whole. For example, when marketing and manufacturing departments have their annual football match, inside the teams the marketing person identify himself with the marketing department – as a sub-system belong to the whole company. But if this same marketing employee talked to strangers of other companies he identified himself as employee of his company – independent in which division, like marketing department, he works: "I am proud that I work for company XYZ". Although the leaders of a company or organization try to promote a shared culture throughout the organization, there are instances of department culture and loyalties being stronger than that of the whole. Such tensions are usual and can be difficult to overcome. Because all persons belong to several and different systems, misunderstandings and conflicts between their various cultures can easily arise. In many instances such conflicts are relatively minor and are worked out instinctively within the system: a kind of self-healing and self-levelling process goes on all the time. But, on other occasions this does not occur some intervention, like support for changing the point of view at an existing working group, that new group members have equal rights - to be a part of the whole (Figure 9.3). In addition to the ambiguities and conflicts caused by the different and incompatible needs from diverse systems, also other questions arise when the systemic principles are not followed.

Figure 9.3. Different relationships in business context

Systemic constellations tend to refer to organizations established for a particular purpose, often that purpose is work/business, but it could also include the army, health, education, religion or a stamp-collecting club. Systemic constellation work is very broad and possible in the context of team-building, supervision coaching process. They work in group settings as well in individual counselling settings. In addition, by using team members in specifically defined ways, a management constellation may help surface and deal with intuitive and other non-rational behavioural drivers in ways that are more sympathetic to existing organizational mindsets and culture, and therefore encounter less resistance notes of business communities (Barber, 2002):

- They can be applied like an instrument of navigation with which move through all kinds of business and organization-related questions.
- They deliver a condensed, 3-dimensional, spatial overview of the status quo, allow possible ways to understand how it came to be and point out options for successful change.
- They illustrate and bring out the questions and issues the client has concerning his company or organization in a simple, precise, and experiential way.
- They offer practical and concrete possibilities for how to proceed.

- They offer alternative perspectives on existing dilemmas and how to transform crisis into chances of growth and transformation.
- They can serve as a simulation for possible solutions (e.g. merging of businesses, strategic marketing changes, organizational changes)

Business proprietors always face challenges within their companies. These challenges can be resolved through systemic constellations alone or in conjunction with modern systems, if required. As an example: sometimes at company multidivisional cooperation's are necessary. In that case modern systems like a digital workflow application are used to prepare multidivisional transparency. But sometimes the systems will leave unaddressed issues, especially if there are problems in the workplace. These issues could address together with systemic constellation work. Furthermore, the modern systems can leave out basic fundamentals that are crucial to a good solution emerging. Examples:

- **Company /Organization Fields**: Repeated training measures don´t bring the desired outcome; motivational problems, high rates of dismissal and employee sickness, problematic working relationships, problems of leadership, definition of a strategy, integration after merger or acquisition of a company, verification of the coherence of the structure of the company, conflict management in the organizations, problematic relations with customers, check-up of task and goal-orientation of the company, test the effectiveness of important steps or changes in a company, identification, re-processing and recycling of dead weight, systemic conditions for buying of companies and fusions, family businesses: identification and clarification of whether disturbances are related to management or to family issues, succession planning, new founding, partnerships.
- **Personal Fields**: Finding your own place in a company, achieving the skills to fill out a function in a company appropriately, find a good place in the company, decision-finding: leaving or staying, conflicts with colleagues or seniors, explaining feelings of excessive demands, burn-out, changing reoccurring behavioural patterns, find the balance between private and professional life.

9.5 Future Challenges - Adaptation of Systematic Constellation in Business

The systemic viewpoint regards organizations and companies as a complex, living, and learning network of relations which are interwoven with each other in many ways. This may be through contracts and official and unofficial hierarchies but also simply through the relationships the employee of a company or organization have with each other (Figure 9.4).

Figure 9.4. Hierarchies/Relationships of the employee of a company or organization

This is one of the reasons why the systemic approach, instead of focusing just on the elements themselves, emphasizes the relationships between elements. By zooming out the picture of the whole so diagnosis becomes more precise and takes the whole into account. When the gaps are identified, the constellation work will fill them in a balanced way for the whole. The following prozess utilization at a later point of time is supposed to support the best in business process with regard to its scope. By creating possible contexts in which win-win situations are more likely to happen, it also invites long-term success. Where for example employees identify with the business they work in motivation and effectively rises. Additional research challenges for constellation work in business are:

- Among peoples and cultures appears a different style. Some cultures are more open to discuss business issues in public workshops and other cultures permit only a limited framework of business issues about public involvement.
- The name *Constellation Work* is used additional by therapy forms. The relationship between therapy and business coaching is not strongly clarified.
- Sometimes in systematic constellation session a mix of private and business conflicts are indicated. In the constellation work it is important to respect and protect the private aspects of the clients – means no open discussion about private clients' aspects in a public group.
- A Management Constellation strongly emphasizes working with issues at the level of tasks rather than focusing on more interpersonal issues. Where interpersonal issues are the cause of disruption, it is possible that a constellation might not be the intervention of choice with the team as a whole (although the highest-ranking team member, for instance, might benefit from an off-line organizational constellation with a group of strangers).

- The introduction of the Management Constellation also requires some skills in using language that is true to the systemic methodology but also grounded and businesslike. The words used to lead the team in to and out of the constellation are important.
- One objective is to minimize the possibility of a change of structural level occurring in the systemic constellation - for example, when a manager's issue is initially located within the organizational domain but suddenly is seen to be much more personal and have its origin in his or her family system. In an open workshop when a traditional organizational constellation is in progress, good practice demands that such changes of structural level are handled sensitively.
- In selecting elements of an issue to be represented, preference is given (as far as is possible) to setting up more abstract elements rather than human elements. For example, in a constellation about a company's marketing strategy the six elements represented included the focus, the agreed marketing approach, that which was outside the agreed marketing approach, the boundary between the two, the market context, and a free element representing an 'x factor'. In a more phenomenological free form organizational constellation at a different time the representatives include the marketing director, the board, the suppliers, customers and the strategy.

9.6 Research Fields - Systemic Constellation Approach

The systemic constellation approach is used in business coaching as group-simulation method (Rosner, 2007). As simulation method it delivers descriptions of cause and effect relationships in organizations and gives the participants an idea of functionality and configurability organizational networks of relationships and rules. It based on a constructivist understanding, defining systemic in a comparative way as the extent to which something (a theory, an explanation or an intervention) relates to the context and relationships in a system. The operational principle of systemic constellation is difficult to explain. Constellation work is a highly systemic method because hypotheses are generated entirely from the relationship dynamics and the interaction between context factors. This is the core understanding of the term systemic, which is defined as example by Matthias Varga von Kibéd as follows:

"An explanation (theory, methodology, approach, definition of terms, thinking, idea, form of therapy, intervention): A is more systemic than explanation (theory...) B, by definition, when A allows a shift away from the attribution of qualities to system elements (and towards the observation of relationships, structures, contexts, dynamics and choreographies) to a greater extent than B" (Kibéd, 2005).

This definition describes systemic in a comparative way. It does not determine 'systemic' in terms of 'Yes' or 'No' but 'more' or 'less, Different perspectives over to explanation models are defined at Weber (Weber *et al.*, 2005) and

Sheldrake (Sheldrake, 2003). In a first attempt constellation work may be considered analogous to the statements quantum mechanical measurement process, there are some striking similarities. For example, a quantum mechanical measurement result is not predictable. Also creates the measurement itself only the result of which was non-existent as before. In addition, the subject-object separation in the quantum mechanical measurement is difficult or impossible, and not only the facts but also measured conditions affect the measurement result. All of this effects will be used in systemic constellation work. Such an analogy is not to be considered its own sake, rather it is to improve our understanding of systemic constellations and to give suggestions for a change in practice in the process of establishing itself. This question is explored in detail in a study that is published in Scribd (Guretzky *et al.*, 2013).

The relevant research in the context of systemic constellation approach is still rare and should be explored with future work in this field. Another attempt in the nature could be clarify the functioning forces in living systems that are otherwise still very challenging to understand and use. There are hints of these same forces in nature when at everyday mysteries such as the synchronized swooping of flocks of birds. If the coordination was based on the birds perceiving each other through their normal senses to understand them, then their change of direction would be staggered looking like dominoes falling. Instead they move as one entity with no delay. The systemic constellations work also relies on modes of communication that we don't yet completely understand. It should help us to move forward and get another point of view (Sheldrake, 2003).

9.7 Conclusion

Systemic constellation work can be applied in the context of businesses, organizations and teams. This approach offers an opportunity to experience and influence the natural balancing forces which impact leaders, teams and whole organizational systems. Through facilitated workshops and systemic constellation interventions it illuminates pressing issues the informal rules and relationships will be transparent, hindering factors are recognized and implicit knowledge in an organization will be make explicit. This allows a new and healing picture to enables the client and their system to move forward more harmoniously, effectively and creatively, to explore the development and innovation opportunities.

In business coaching and consultant the systematic constellations approach is used to focus on restructuring, communication, team motivation, ethical questions, decision-making, strategic issues and mission statements. Systemic constellations may also reveal a simple and memorable picture of the dynamic that allows the organization to have a compassionate overview of the entire situation.

9.8 References

Barber, P. (2002) Gestalt in Coaching & Consulting: A Dialogue with Holism & The Soul, Organizations & People

Bodirsky, Chr. (2013) Systemische Aufstellungen leiten – Therapie und Praxis, Institut SWT, Hannover, Chapter 1 Basics education edition not published

Clutterbuck, D., Megginson, D. (2009) Further Techniques for Coaching and Mentoring, Elsevier Butterworth Heinemann

Guretzky, B. v., Hölscher, Th., Pape, K. (2013), Quantenmechanischer Messprozess und Strukturaufstellungen published at http://de.scribd.com/doc/154296458/Quanten mechanischer-Messprozess-und-Strukturaufstellungen-pdf

Hellinger, B., Weber, G., Beaumont H. (1998) Love's Hidden Symmetry; What Makes Love Work in Relationships, ISBN 978-1-891944-00-0

Migge, B (2011) Handbuch Business Coaching, Beltz Weinheim, Basel 2011, ISBN: 978-3-407-36463-0

Rosner, S. (2007) Systemaufstellung als Aktionsforschung – Grundlagen, Handlungsfelder, Perspektiven, Verlag Dr. Rainer Hampp München und Mering, ISBN 978-3-86618-150-2

Sheldrake, R. (2003) The sense of being stared at and other aspects of the extended mind., pp 113-116. Random House London

Sparrer, I. (2009) Systemische Strukturaufstellen, Auer-System-Verlag

Sparrer, I. (2011) Wunder, Lösung und Systeme. Lösungsfokussierte systemische Strukturaufstellungen für Therapie und Organisationsberatung, Carl-Auer-Systeme Verlag, ISBN 3-89670-2041

Varga von Kibéd, M. (2006) Solution-Focused Transferability: How to keep the Essence of the Solution-Focused Approach by extending it. In: Lueger und Korn (Hrsg.) Solution-Focused Management, Band 1, Rainer Hampp Verlag, München und Mering, ISBN 978-86618-098-7

Weber, G. (2004) Organisational Constellations: Basics and Special Situations, translated by Peterson, J. & Luppertz, U. from Praxis der Organisationsaufstellungen, Carl Auer Systeme Verlag

Weber, G., Schmidt, G., Simon, F. B. (2005) Aufstellungsarbeit revisited … nach Hellinger? Carl-Auer Verlag, ISBN: 978-3-89670-486-3

Chapter 10

Process Indicators for Process Engineering (PIPE)

M. Schabacker and M. Gröpper

10.1 Introduction

The campaign Process Indicators for Product Engineering (PIPE) of the companies CONTACT Software, Dassault Systèmes, IBM Software Group, IBM Global Business Services, and Transcat PLM Germany has taken over on the task of evaluating Product Lifecycle Management (PLM) solutions using standardised indicators for the product engineering process. This campaign was supported by VDMA (German Engineering Federation) as a non-materialistic sponsorship and by the Chair of Information Technologies in Mechanical Engineering, Otto-von-Guericke University Magdeburg. The chair provided scientific consulting and monitoring. After studying the literature there was only one reference publication (Alemanni *et al.*, 2008) devoted to the subject of indicators in Product Lifecycle Management. These authors had collected a lot of number indicators (e.g. part list changes number, change issue number). The objective of the PIPE campaign was to create process indicators with an interpretation context through feature cluster analysis of the product life cycle phases, indicating weaknesses in products, processes or business units, or through evaluation of formulas.

10.2 Evaluation Problem of Engineering Processes

In recent years, efforts were at the forefront of predicting the benefit aspects of PLM solutions prior to their introduction and their quantification in monetary terms. Today, the evaluation of PLM solutions after their introduction by characteristic factors is in the focus of attention.

Characteristic factors provide complex subjects in a simple (e.g. the number of engineering change requests in product development) and compressed form (e.g. early warning rate of redundant engineering change request). An *indicator* is a

M. Schabacker et al. (eds.), *Modelling and Management of Engineering Processes*, DOI 10.1007/978-3-662-44009-4_10

characteristic factor that can be interpreted by the user and is based on predefined values (target-actual comparison or target value). The indicator may relate to performance or capacity of the company, its individual organisational units or a machine (Schabacker and Simon, 2012). Therefore, many meaningful indicators have mainly been shown for machinery and equipment in the manufacturing. They were merged for example in the VDMA 66412-1 (VDMA 66412, 2008) and ISO 22400-2:2014 (ISO 22400-2, 2014).

But how can the definition of the indicator be applied to the environment of product development? For example in manufacturing processes, the indicator staff productivity (ratio of productive time to the present time) is fully justified. However, it is pointless in product development to regard the productive work (e.g. to measure the number of created drawings or 3D CAD models per day). There are three important influences within product development:

- the ever-changing customer requirements, which lead only to engineering change requests and after this (economic) evaluation into engineering change orders,
- easy configurable products so that the customers get exactly those products they like,
- effectively organised product development projects in project teams that develop new and complex solutions in less time.

A further influence requires the cooperation of the component-level products between internal and external product development partners to cause comprehensible decisions as quickly as possible.

Therefore, and because of these influences, the project partners of PIPE have developed indicators for the cross-cutting processes in Product Lifecycle Management:

- change management
- requirements management
- configuration management
- project management
- collaboration management

These so-called PIPE processes are described in section 10.3.

10.3 PIPE Processes

Figure 10.1 shows the four parallelly running PIPE processes *change*, *requirements*, *configuration*, and *project management* as well as *collaboration management* as a cross-cutting PIPE process compared to the other four PIPE processes.

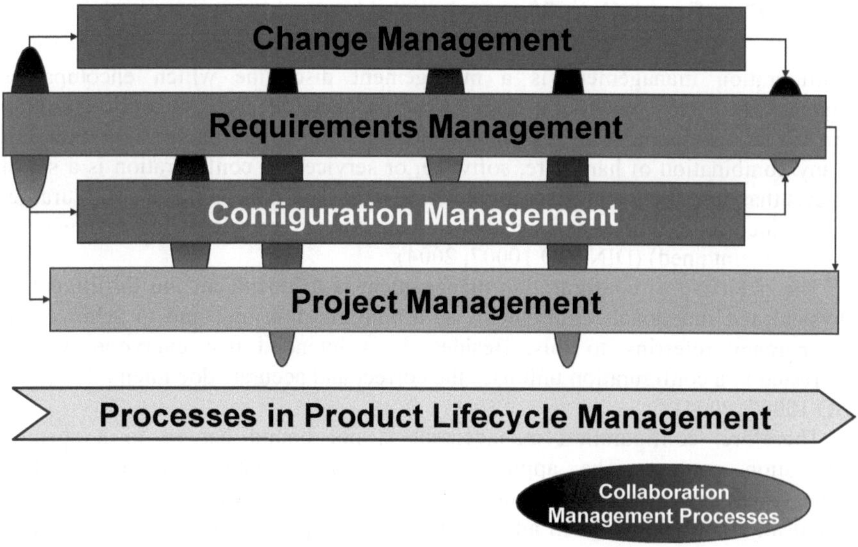

Figure 10.1. Flow of the PIPE processes in Product Lifecycle Management

10.3.1 Change Management

The change management supports all activities for the acquiring, collecting, evaluating, deciding, planning, and adding of product changes. Change management includes five sub-processes:

- identifying the reason for change
- specifying the engineering change request
- evaluating the engineering change request in the departments
- bringing together reviews and documenting decisions
- evaluating change alternatives with defining milestones

10.3.2 Requirements Management

Requirements management supports target-oriented development of products which meet specifications. It involves collecting, processing , structuring, and verifying of customer and internal requirements. Requirements management includes five sub-processes:

- collecting customer and internal requirements
- specifying requirements
- structuring and classifying requirements
- monitoring the implementation of the requirements
- reviewing requirements after the implementation

10.3.3 Configuration Management

Configuration management is a management discipline which encompasses organisational and behavioural rules to the product life cycle of a configuration unit via its development through manufacturing and support. A configuration unit is any combination of hardware, software, or services. A configuration is a set of objects that describe the product for a certain time. The status of each configuration are retained as baselines and can be built up for several views (as designed, as built, as maintained) (DIN ISO 10007, 2004).

The objective of configuration management is to document the fulfillment of physical and functional requirements of a configuration unit and to achieve full transparency referring to this. Besides, it is intended that everyone who is interested in a configuration unit uses the correct and accurate documentation (DIN ISO 10007, 2004).

Therefore, configuration management is not bound per se to a specific application context. The application and implementation of configuration management result in a configuration management process. This requires the organisation and planning. In addition to this conceptual section of configuration management four sub-processes of configuration management are to be distinguished (DIN ISO 10007, 2004):

- identifying the configuration(s)
- executing the documentation in configuration management (the so-called configuration record keeping)
- monitoring the configuration(s)
- executing the configuration auditing

10.3.4 Project Management

Project management comprises, in accordance with DIN 69901 (DIN 69901, 2009), the entirety of managerial functions, organisation, techniques, and means for the execution of a project. The project management includes four sub-processes:

- initialising the project
- planning the project
- monitoring the project
- completing the project

10.3.5 Collaboration Management

Collaboration Management controls the collaboration on component-level between internal and external partners to bring about as quickly as possible comprehensible decisions. Collaboration Management is a cross-section process to change, requirements, configuration, and project management.

10.4 PIPE Indicators

To assist companies in the identification of optimisation potentials in the product creation process using standardised process indicators, the focus of this campaign is therefore based on the Continuous Improvement Process (Schabacker and Simon, 2012). This can be reached by reorganisation, improved processes and methods as well as tools (IT systems such as CAx or PDM systems) in product development. Methods to improve training can also make a valuable contribution because for example unified modeling methods at the 3D CAD workstation improve the cooperation. A process indicator supports in

- making decisions (problem identification, representation, information extraction),
- the control of processes (target-actual comparison),
- the documentation of processes (traceability and transparency of processes) and/or
- coordination (behaviour control) of important facts and relationships in the product creation process of a company.

10.4.1 Description of the Process Indicators

The description of the process indicators is based on the VDMA 66412-1 (VDMA 66412, 2008) which includes characteristic factors for Manufacturing Execution Systems (MES).

The data for a process indicator can be collected for the actual recording either manually or read from existing IT systems (e.g. PDM systems) in a time period to be determined beforehand (monthly, quarterly, fiscal year, per project). The process indicators are designed so that the two cases are considered. Some process indicators include correlations of variables formula (indicated by the formulation for example ...ratio, ...degree), others include values based on an accumulated number of features or feature clusters (e.g. reason for change, product component, corporate responsibility). Besides, a process indicator includes

- the product creation phase (e.g. pre-design, design, manufacturing, production support),
- its benefits and/or identification of improvement potentials and/or application,
- evaluation using a trend statement (e.g. the higher, the better) and further evaluation instructions and explanations as well as
- its mapping to corporate responsibilities or departments of a company or corporate level managers, the indicator relates to this, and
- its mapping to the PIPE processes.

An example of the description of a process indicator is shown in Figure 10.2.

Name of the Indicator:	Early Warning Rate of Redundant Engineering Change Requests
Description of the Indicator	
Phase of the Product Creation Process:	Unnecessary process steps sooner redundant requests are detected, the sooner and avoided other processes continue or be terminated.
Benefit/ Potential for Improvement Application:	The proportion of early detected redundant requests should tend to increase due to the learning curve over time.
Time Behaviour:	continuous
Definition and Calculation of the Indicator	
Formula:	$NoECR_{RED\ err\ pre}$ = Number of Engineering Change Requests$_{RED\ err\ pre}$ $NoECR_{RED\ err}$ = Number of Engineering Change Requests$_{RED\ err}$ $EWRoECR_{RED\ err}$ = Early Warning Rate of Engineering Change Requests$_{RED\ err}$ $EWRoECR_{RED\ err} = \frac{NoECR_{RED\ err\ pre}}{NoECR_{RED\ err}} * 100\%$
Unit/ Dimension:	%
Evaluation:	Trend: the higher, the better
Analysis:	Due to a learning curve, the proportion of early warnings should continuously improve. The smaller the percentage, the greater the improvement in rate should be. Early detection rate close to 100% can only be seriously improved.
Notes	
Hints/ Explanations:	-
Corporate Responsibility:	☑ Marketing ☑ Development/Design ☐ Manufacturing ☐ Service ☐ Purchase
Company Level:	Middle management: development management / design management, portfolio management, product managers, controlling
PIPE Process:	☑ Change Management ☐ Requirements Management ☐ Configuration Management ☐ Project Management ☐ Collaboration Management

Figure 10.2. Sample description of a process indicator

All process indicators are described in the VDMA guideline *Process Indicators for Product Engineering (VDMA, 2013).* Table 10.1 shows 24 process indicators for the five PIPE processes.

Table 10.1. Process indicators for change management

Name of PIPE Process	Name of Process Indicator
1. Change Management	1.1 Number of Engineering Change Requests (ECR) 1.2 Number of Engineering Change Orders (ECO) 1.3 Abnormality Degree (change cluster) 1.4 Throughput Time of ECR 1.5 Throughput Time of ECO 1.6 Early Warning Rate of Redundant ECR
2. Requirements Management	2.1 Approval Rate of Requirements 2.2 Processing Status of Requirements 2.3 Degree of Coverage of Acceptance Criteria
3. Configuration Management	3.1 Carry-Over of Parts Degree of Standard Parts 3.2 Carry-Over of Parts Degree of Product Components 3.3 Configuration Items Ratio 3.4 Structured Configuration Items Ratio 3.5 CAD Documents Configuration Items Ratio 3.6 Historicised Configuration Items Ratio 3.7 Synchronised Configuration Item Ratio
4. Project Management	4.1 Adherence to Delivery Dates 4.2 Cost Efficiency 4.3 Project Status 4.4 Number of Duration of Missed Deadlines (internal/external partner)
5. Collaboration Management	5.1 Knowledge Management Degree of Maturity 5.2 Automation Level of Processes supported by automated Workflows 5.3 Partner Evaluation Status 5.4 Part of Search Time per Employee

Table 10.2 shows the process indicators for requirements management with their variables and formulas. Two of these process indicators have one interpretation context and reporting option, respectively; the process indicator *Degree of Coverage of Acceptance Criteria* has two reporting options. The process indicators for the other four PIPE processes are described in a similar way. In total, the campaign developed 24 process indicators with 37 reporting options.

Table 10.2. Process indicators for requirements management

Name of Process Indicator	Name of Variables with Formula Context
2.1 Approval Rate of Requirements	*NoACR* = Number of Approved Consistent Requirements *NoR* = Number of Requirements *ARoR* = Approval Rate of Requirements $ARoR = \frac{NoACR}{NoR} * 100\%$
2.2 Processing Status of Requirements	*NoRR* = Number of Raw Requirements *NoR* = Number of Requirements *DoCoRwS* = Degree of Coverage of Requirements with Solutions $DoCoRwS = \frac{NoRR}{NoR} * 100\%$
2.3 Degree of Coverage of Acceptance Criteria	*NoDAC* = Number of Defined Acceptance Criteria *NoRR* = Number of Reworked Requirements *NoR* = Number of Requirements $DoCoAC_{complete}$ = Degree of Coverage of Acceptance $\text{Criteria}_{complete}$ $DoCoAC_{range}$ = Degree of Coverage of Acceptance Criteria_{range} • Completeness of Acceptance Criteria: $DoCoAC_{complete} = \frac{NoDAC}{NoR} * 100\%$ • Scope of Post of Requirements: $DoCoAC_{range} = \frac{NoRR}{NoR} * 100\%$

10.4.2 Evaluation of the Process Indicators

To demonstrate the feasibility of the results the project partner developed a PIPE tool based on Microsoft Excel.

One possibility for the visualisation of results is the use of evaluation graphs over the entire product lifecycle for the process indicators in change and project management. For example, by summing up the change requests, an accumulation can be detected in a cluster analysis of the product life cycle phases, indicating weaknesses in products, processes or business units. This is used with the Pivot method using tables as table data in different cases can be represented and analysed in a condensed, summarised form (Figure 10.3).

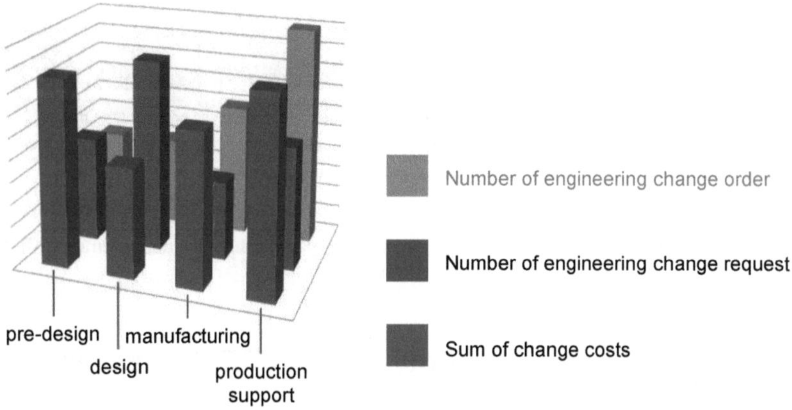

Figure 10.3. Cluster analysis of change management with Pivot tables (Schabacker and Simon, 2012; Schabacker, 2013)

The PIPE tool also allows visual evaluation of the process indicators using a traffic light system. After the company-specific values (target-actual values) are summarised and defined for each process indicator, the comparison with the results for the respective process indicator is carried out:

- The traffic light color *green* indicates that the calculated value for the corporate responsibility (responsibilities) and the phases of the product creation process moves within the permissible range, so no continuous improvement must necessarily take place in the process.
- The traffic light color *yellow* indicates that the calculated value is **within** a tolerance range outside the company-specific value for the corporate responsibility (responsibilities) and the phases of the product creation process; here, a continuous improvement for the respective corporate responsibility and respective phases of the product creation process should be planned.
- The traffic light color *red* indicates that the calculated value is **far beyond** the tolerance range of company-specific value for the company responsibility (responsibilities) and the phases of the product creation process; here it must be planned a continuous improvement for the respective corporate responsibility and respective phases of the product creation process.

An example for an evaluation is shown in Figure 10.4 for requirements management (for the calulation see formulas in Table 10.2).

Result for PIPE Indicators
Period
Begin: 31.01.13
End: 15.03.14
Period Length: 408 days

PIPE Process	PIPE Indicator	Result	Company-specific Values	Unit
Requirements Management				
	2.1 Release Degree of Requirements	80,00%	70%	%
	2.2 Processing Status of Requirements	80,00%	82%	%
	2.3 Degree of Coverage of Acceptance Criteria	20,00%	70%	%
		30,00%	70%	%

Figure 10.4. Process indicators for the evaluation of requirements management (VDMA, 2013)

10.4.3 Evaluation of Maturity of the PIPE Processes

An overall evaluation of all process indicators in the PIPE processes is based on the achieved number of traffic light colors *green*, *yellow*, and *red* (Figure 10.5).

PIPE Compass	Number	Number	Number	Overall Evaluation Indicators	Maturity
Change Management	6	2	5	**13**	**53,8%**
Requirements Management	1	1	2	**4**	**37,5%**
Configuration Management	0	1	6	**7**	**7,1%**
Project Management	5	0	3	**8**	**62,5%**
Collaboration Management	2	0	3	**5**	**40,0%**
Overall Traffic Analysis	**14**	**4**	**19**	**37**	**43,2%**

Figure 10.5. Example of the evaluation of the traffic light functions for all process indicators and PIPE processes (VDMA, 2013)

A maturity level is calculated based on the number of traffic light colors of indicators for each PIPE process and represented in a spider web diagram (Figure 10.6).

This reflects the current state of a company. However, this does not mean that, irrespective of the costs, a 100% state (i.e. company-specific values are exceeded at all process indicators) must be achieved by using the methods of the continuous improvement process.

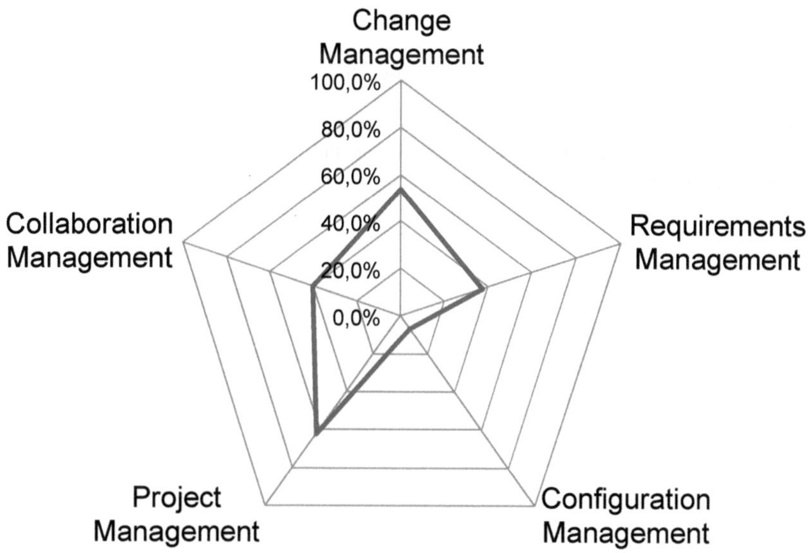

Figure 10.6. Example of the maturity level for all PIPE processes (VDMA, 2013)

10.5 Conclusions

After the successful project completion of this campaign 24 process indicators with 37 reporting options are available in the VDMA guideline *Process Indicators for Product Engineering (PIPE)* that consider all relevant PLM disciplines and are already being used by customers. The benefit of PIPE lies in a uniform evaluation methodology. The identified process indicators have been tested with companies using the PIPE tool with respect to their applicability and feasibility.

10.6 References

Alemanni M., Alessia G., Tornincasa S., Vezzetti E. (2008) Key performance indicators for PLM benefits evaluation: The Alcatel Alenia Space case study, Computers in Industry 59 pp 833-841

DIN ISO 10007:2004-12 (2004) Qualitätsmanagement – Leitfaden für Konfigurationsmanagement, Beuth Verlag, Berlin

DIN 69901 (2009) Projektmanagement – Projektmanagementsysteme, Beuth Verlag, Berlin

ISO 22400-2:2014 (2014) Automation systems and integration – Key performance indicators (KPIs) for manufacturing operations management – Definitions and descriptions

Schabacker, M., Simon, A. (2012) Qualitätsmanagement für die Produktentstehung – Entwicklungsprozesse messbar machen, in: IT&Production PLM Wissen Kompakt, 2012/13, pp 32-34
Schabacker, M. (2013) Engineering-Prozesse einheitlich bewerten, in: SPS-Magazin Ausgabe 7/2013, pp 24-25
VDMA 66412-1 (2008) Einheitsblatt MES-Kennzahlen Teil 1, VDMA-Verlag, Frankfurt
VDMA (2013) Leitfaden Process Indicators for Product Engineering (PIPE), VDMA-Verlag, Frankfurt

Chapter 11

Comparison of Seven Company-Specific Engineering Change Processes

M. Wickel, N. Chucholowski, F. Behncke and U. Lindemann

The management of engineering changes is an ongoing topic in academia and practice. To define the ideal engineering change process is still a challenge due to the opacity for necessary activities and the lack of efficacious supporting methods and tools. The comparison of seven company-specific engineering change processes gives insights into a detailed activity level of engineering changes in practice. By comparing the processes based on a reference process, commonalities and differences are derived. Coincidental, a generic engineering change reference process was developed, which describes an ideal process with all possible activities and process steps when dealing with engineering changes.

11.1 Introduction

Engineering changes absorb up to 31% of the product development capacity (Maier and Langer, 2011). This stresses the importance of their management, which is still an ongoing topic in academia and practice. One of the frequently upcoming topics is the opacity of engineering change processes (ECPs) and all necessary activities when dealing with engineering changes (ECs). Even if this was already a research topic several years ago (Eckert *et al.*, 2004; Jarratt and Clarkson, 2005; Lindemann and Reichwald, 1998; Terwiesch and Loch, 1999) and industry is aware of the proposed ECPs, there are still a lot of problems when dealing with ECs in practice. The described ECPs in literature are either not specific enough to use them, or are too specific what makes it difficult to adapt them to companies' needs in ECM (Jarratt and Clarkson, 2005).

In order to derive best practices and to enhance the management of ECs in industry, a comparison of company-specific ECPs was conducted. To allow a consistent comparison, there is a need for a reference process. The development of such an initial engineering change reference process (ECRP) is described in section

M. Schabacker et al. (eds.), *Modelling and Management of Engineering Processes*, DOI 10.1007/978-3-662-44009-4_11

11.4, before the comparison of seven company-specific ECPs is addressed in section 11.5. This includes a revision of our ECRP, the description of the comparison and a discussion of main findings. Section 11.6 concludes the chapter and gives an outlook for further research. The following section gives an overview on the way the research was conducted and describes the corresponding research methodology.

11.2 Research Methodology

Following the strong interest by practitioners in a comparison of ECPs, we conducted a workshop on that topic within an expert group. In order to provide a common basis for terms and understanding of ECPs, we developed a reference process model for engineering changes based on literature. Thereby, we considered processes presented in commonly used scholar databases, engineering standards as well as book publications on engineering change management. The resulting model was used by every workshop participant to describe the ECP within their company preliminary to the meeting. In the meeting, the ECPs were discussed and commonalities and differences were identified in a workshop. Afterwards we adapted the reference process based on the findings in order to provide a generic framework for a comparison of company-specific ECPs and summarized the findings of the comparison in the new framework. In another meeting we evaluated the new ECRP and the findings of the comparison together with further participants of the industrial working group. The participants of the meetings and our research approach are described in more detail in the following sections.

11.2.1 Description of Participants

The group of experts on engineering change management was founded in 2012 and meets three times a year. The group was founded following the interest in more opportunities for discussion and knowledge transfer about ECM between academia and practice. Usually, five to ten participants from different companies attend the meetings, discuss current challenges and draw the advancements of ECM in academia and practice based on the fields of action in industry. The companies represented reach from middle-sized enterprises to large-scale enterprises and from suppliers to original equipment manufacturers (OEMs).

A total of 13 practitioners from seven different companies participated in the first meeting. The objective of this meeting was a comparison of ECPs within the different companies. Anonymized details about the represented companies among the participants are provided in Table 11.1.

Table 11.1. Information about participants of the industrial working group

Company size (employees)	Position in supply chain	Positions of representatives	Industry
~80.000	OEM	Change manager	Commercial vehicles
~45.000	OEM	Process manager	Home appliances
~35.000	OEM	Change manager, process manager	Commercial vehicles
~20.000	OEM	Project/Change manager	Fixation systems
~17.000	OEM	Change manager	Commercial vehicles
~15.000	Supplier	Change manager	Automotive
~1.200	Supplier	Head of development	Manufacturing

11.2.2 Research Approach

In order to compare company-specific process steps in ECM, a reference process was needed as a basic model for comparison. In a first step, we developed an initial EC reference process model based on literature about ECPs. We used this model for the comparison of ECPs of the participants, who allocated their company-specific processes within our model. Together with the practitioners, we discussed the different processes in a workshop and reorganized the different company-specific process steps. Then, we analyzed the workshop results in order to derive commonalities between the companies and to identify gaps; on the one hand, that might be process steps in our initial ECRP which are not used in industry or one the other hand process steps which are not captured in our initial ECRP. As a consequence, we restructured our reference process and changed the level of detail for some activities within the process. As a last step, we presented the resulting ECRP within a second meeting of the industrial working group with further participants, including the allocation of their company-specific ECPs.

11.3 Development of the Initial ECRP

As described before, we developed a reference process model based on literature in order to provide a framework for the comparison. The following sections give an overview of literature that deals with engineering change processes and describes how we derived our initial engineering change reference process (iECRP).

11.3.1 Perspectives on Engineering Change Processes in Literature

As defined by Jarratt *et al.* (2011) an EC is an alteration made to a product or its documentation. Furthermore, they define the organization and controlling of the processes for an EC as Engineering Change Management (ECM). The process behind the management of ECs is often called ECM process or just EC process

(ECP) (Jarratt *et al.*, 2011). The EC process has similarities with the conventional design process or problem solving processes in general as presented by e.g. Pahl *et al.* (2007): confrontation, information, definition, creation, evaluation and decision. But there are also some important differences. Aßmann (2000) mentions inter alia:

- The main focus of engineering processes is the generation of data. In EC processes existing and shared data are modified;
- EC processes are characterized by a variety of administrative steps in order to minimize potential side effects of ECs;
- Due to a large number of boundary conditions (e.g. existing data, increased pressure of time) is the planning and coordination of ECs within design processes complex.

The statements about which processes exactly are part of this ECP differ in literature significantly. The different perspectives on ECM in literature are listed in Table 11.2.

Table 11.2. List of different understandings of ECM in literature.

Characterizations of ECM	Reference
ECM usually includes the four stages: Identifying; Evaluating; Implementing; Auditing Moreover, some common activities: identification and control of product structures; maintenance of revision control; history of all changes of products and its associated documents.	(Huang and Mak, 1999)
ECM is the process of making engineering changes to a product in a planned or systematic way, including the following steps: • Emergence of a need for the change • Request for the change • Management approval of the change • Implementation of the change • Documentation of all impacted product data	(Rouibah and Caskey, 2003)
• Engineering change request raised • Identification of possible solution(s) of change request • Risk/Impact assessment of possible solution(s) • Selection and approval of a solution by change board • Implementation of solution • Review of particular change process	(Jarratt and Clarkson, 2005)
ECM encompasses all documents, methods, actions and processes that are necessary for the avoidance, anticipation, effective selection, processing, approval/disapproval, execution, control and documentation of engineering changes.	(Köhler, 2009)
• Identify change: Initiate problem, Estimate problem, Request change, Initiate Solution • Propose change: Analyze and order change, Propose solution • Alteration: Plan change, verify plan, execute and approve • Implementation of change: Estimate impact, Release change, Modify orders/requests/configuration, Disclose change	(Rozenfeld *et al.*, 2009)
• Clarification of the change case • Selection of change mechanism(s) • Evaluation of alternative change options • Actual decision-making and approval of a change option • Implementation • Review of the individual change process and lessons learned	(Kissel and Lindemann, 2013)

Most of the authors describe ECM as the processes and actions to handle ECs after the need for them was already identified (i.e. the starting point is an EC request). Obviously, everybody mentions the implementation of the change itself as part of ECM. The generation of possible solutions, a risk and impact analysis and the decision process in advance of the implementation, and the retrospective review of already executed changes are not always included.

In order to list all necessary process steps of ECM it is also useful to take a look at the strategies pursued by ECM. Table 11.3 summarizes the different strategies for ECM described in literature. Most of the strategies can be found implemented in the different process steps presented before. Only the avoidance, reduction and the front-loading of engineering changes cannot be matched to the identified process steps within ECM. This is due to the different characteristic of these activities. The ECPs described in literature are executed every time when there is a target deviation. The activities to avoid, reduce or anticipate changes take place on another level, i.e. are incorporated within the overall development process (Lindemann and Reichwald, 1998).

Table 11.3. Summary of ECM strategies mentioned in literature.

Strategies	References
Avoid and reduce engineering changes	(Lindemann and Reichwald, 1998; Terwiesch and Loch, 1999; Aßmann, 2000; Fricke *et al.*, 2000; Rouibah and Caskey, 2003; Eckert *et al.*, 2004)
Front-loading of engineering changes	(Terwiesch and Loch, 1999; Aßmann, 2000; Fricke *et al.*, 2000; Rouibah and Caskey, 2003)
Effective and fast decision making on change implementation	(Fricke *et al.*, 2000; Rouibah and Caskey, 2003; Jarratt *et al.*, 2011)
Reduce negative impact of engineering changes	(Terwiesch and Loch, 1999);
Efficient implementation of engineering changes	(Terwiesch and Loch, 1999; Aßmann, 2000; Fricke *et al.*, 2000; Jarratt *et al.*, 2011)
Learning from previous engineering changes	(Fricke *et al.*, 2000)

11.3.2 The Initial Engineering Change Reference Process (iECRP)

By merging all different strategies and process steps identified in literature into one process model, a basis for the comparison was derived. The result is a model (iECRP) with five phases:

- Identification of the necessity for a change;
- Preparation of the change (generation of options and their assessment);
- Decision for a change option;
- Operation: implementation of the change;
- Review of change effects.

These phases are on a too abstract level in order to allocate concrete activities within the handling of ECs in industry. Figure 11.1 shows the initial ECRP where every phase is detailed with necessary activities in an ideal procedure of an EC. Since this model was revised in a next step, the detailed description of the phases follows in the next section.

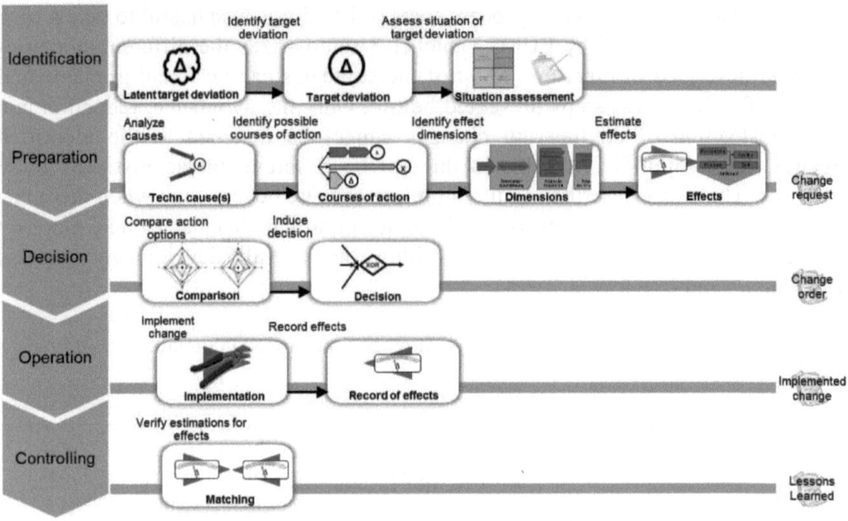

Figure 11.1. Initial model for an engineering change reference process.

11.4 Comparison of Seven Company-Specific Engineering Change Processes

11.4.1 Revision of the ECRP According to the Company-Specific EC Processes

The iECRP based on literature (see Figure 11.1.) was sent preliminary to the meeting to the participants of the industry working group with a description of the iECRP. The representatives of the seven companies then allocated their company-specific ECP to the iECRP in preparation for the working group meeting.

Within the workshop the participants presented consecutively their company-specific ECPs with reference to the phases and process steps of the iECRP. Subsequently a discussion was lead about differences and commonalities of the seven company-specific ECPs.

The allocation of the company-specific EC processes to the iECRP indicated that some revisions would be helpful to reach a better result in the comparison afterwards. Therefore the following points were revised:

- simplification to a purely activity-oriented process model (the results of the activities are depicted by the graphics);
- aggregation of two activities ("Identify effect dimensions" and "estimate effects" to "Identify and estimate effects");
- expansion of the activity "verify estimations for effects" to "match results";
- modelling "Lessons Learned" as a process activity: "draw lessons learned".

Figure 11.2 depicts the revised ECRP with five phases and eleven process activities. It is of prime importance that the whole process should not be seen as just sequential. There are many loops possible and the sequence of actions depends on the specific context and situation, which are investigated in the identification phase.

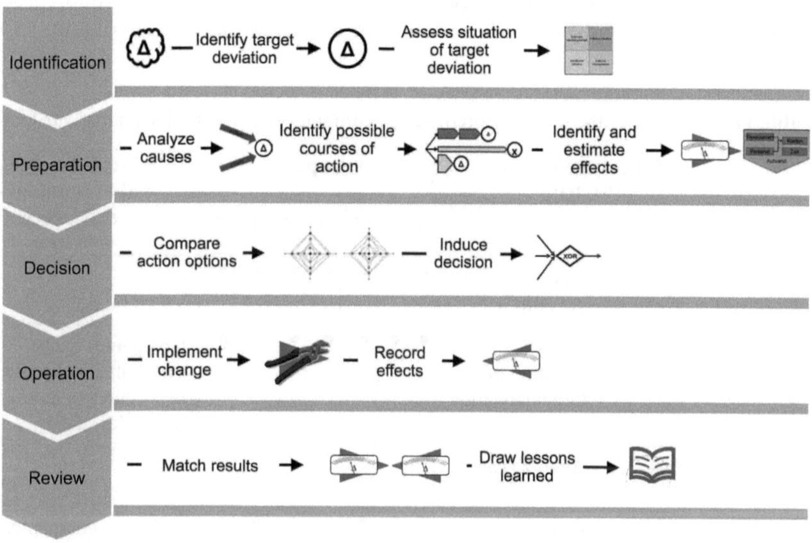

Figure 11.2. Revised ECRP according to the company-specific EC processes

The identification of the target deviation and the assessment of the situation regarding the necessity to change and boundary conditions are focus of the first phase (identification). Here, the further procedure is defined roughly. Within the following preparation phase, more information is gathered. Among with a cause analysis in order to identify the technical cause behind the target deviation, several courses of action are elaborated. For each course of action, the dimensions and extents of resulting effects are estimated. The result of the preparation phase is an engineering change request (ECR) that describes the target deviation, the underlying technical cause, potential solutions and related effects. Based on the ECR, a decision has to be made in the next phase whether there will be a change at all and if so, what solution should be implemented. The result of the decision phase is an engineering change order (ECO). During the operation phase, all actions described in the ECO are executed. The effects of the change have to be recorded during implementation in order to derive lessons learned in the closing controlling phase, where estimated change effects are compared to actual effects. Hence, the quality of predictions for change effects can be enhanced. The following section presents the comparison of the company-specific ECPs, allocated in our ECRP.

11.4.2 Comparison of Company-Specific ECPs

The seven company-specific ECPs were compared against the revised ECRP which consists of five phases and eleven process activities. The ECRP is represented in Table 11.4. In a first step of the comparison it was analyzed whether the particular process activities are part of the companies' ECP (see Table 11.4, column: "Quantity of companies"). In a second step the core differences within process activities of particular ECPs were determined (see Table 11.4, column: "Core differences").

Table 11.4. Results of the comparison of seven company-specific ECPs against our ECRP

	Process activities (ECRP)		**Company**							**Core differences within company-specific process activites**
			1	**2**	**3**	**4**	**5**	**6**	**7**	
Identification		Identify target deviations	●	●	●	●	●	●	●	-
		Assess situation of target deviation	●	●	●	●	●	●	●	Companies have different criteria and procedures to assess target deviations.
Preparation	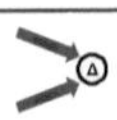	Analyze causes	○	○	○	○	○	○	●	-
	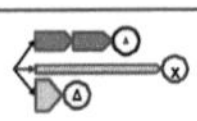	Identify possible courses of action	●	●	●	●	●	●	●	Only two companies generate more than one course of action.
	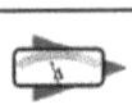	Identify and estimate effects	●	●	●	●	●	●	●	Companies take different effects into accout, which have to be estimated.
Decision	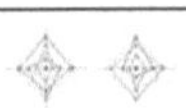	Compare action options	●	●	●	●	●	●	●	Two companies decide between alternatives. The others have "go/no-go" decisions.
	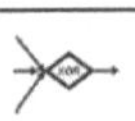	Induce decision	●	●	●	●	●	●	●	Single- or multi-stage decisions are possible.
Operation		Implement change	●	●	●	●	●	●	●	"Just do it" – The activity depends on the

	Process activities (ECRP)		Company							Core differences within company-specific process activites
			1	2	3	4	5	6	7	
										specific EC and company surroundings.
		Record effects	○	○	○	○	○	○	○	-
Controlling		Match results	○	●	●	○	●	○	●	Different conditions are matched by companies: - objectives achieved? - estimations correct? - assumptions occurred?
		Draw lessons learned	●	○	●	○	○	○	○	-

Legend: • Company with process activity ○ Company without process activity

11.4.3 Findings and Discussion

The comparison of the seven company-specific ECPs indicated that the "identification phase" is very important for ECPs and especially the phases "preparation" and "controlling" differ strongly within the seven companies and between academia and practice.

Within the **"identification phase"** the following process activities are determined for the specific situation of the change. Therefore, first the situation and deviation is assessed by the companies. Hereby the point of time within the development process when the deviation is detected is very important. Three out of seven companies differentiate between changes occuring in the planning phase, in the development or production phase or in the phase of product care. Furthermore, one company takes into account if the change affects a complex product or process and also if the customer has to be informed.

The **"preparation phase"** in which causes for target-deviations have to be identified as well as possible courses of action is less emphasized within the companies. Only one out of seven companies has "analyze causes" as a process activity in their ECP and only two companies generate more than one course of action in order to eliminate the target-deviation. Due to increasing time pressure, companies often detail, assess and document only one course of action, which is then captured in a change request. In the following decision phase it will be decided whether the change request will be implemented or not. The decision then is a "go/no-go" decision instead of a decision between alternatives. In discussion with participants of the working group about the preparation phase it becomes apparent that the companies are not sure which effort is appropriate for the preparation phase, i.e. how much time they are allowed to spent in order to find a solution to eliminate the target-deviation. In the foreground there was lead a discussion about the effort-benefit ratio. Besides that, none of the companies documents alternative courses of actions, which were established in the preparation phase. They are not captured in the change request. The ECRP includes this activity and the working group agreed on the potential of this process activity.

Furthermore the **"controlling phase"** differs strongly between the companies and the established ECRP. Within the process activity "match results" the EC is examined retrospectively: two companies check whether the objectives of the EC were achieved and therefore the target-deviation was eliminated, one company checks whether the estimations have been correct and one company checks whether the proposed assumptions have become true (e.g. sales volume for a product). Then only two out of seven companies draw "Lessons Learned" after a closed EC to generate knowledge out of the findings and preserve the knowledge that has been gained during the ECP. The representatives of this two companies which draw lessons learned admitted that they draw lessons learned for some critical ECs but without a structured documentation and procedure and also the distribution of the gained knowledge is not organized.

With regard to the **whole EC process** and the quantities of companies which perform the particular process activities it can be assumed that most of the companies focus on activities which are really necessary to implement ECs. Activities which do not lead or contribute directly to an elimination of the target-deviation or have any benefit for the specific EC are not part of the ECP. Therefore often only one course of action is prepared, assessed and documented within the change request, which leads then to an easy "go/no-go" decision but not necessarily to the best possible solution. Furthermore, no profound controlling is done after a change is implemented so that it cannot be assessed whether the decision was right as well as all assumptions which have been made during the process. Companies thereby abstain from the strategy "learning" out of ECs.

11.5 Conclusion and Outlook

Literature provides numerous similar ECPs on a very high and abstract level. However, these processes lack of detail to apply them as a reference process in order to compare company-specific EC processes. So within this work first an ECRP was developed which is more detailed and profound then the processes already presented in literature.

This ECRP was the basis for a subsequent comparison of seven company-specific EC processes to determine the state of EC processes in industry. The focus within this study was on standard process activities and their differences.

The main findings are that most of the companies generate, document and assess only one course of action instead of several in preparation for an EC decision. Furthermore, the process activities after an EC was already implemented are very poor emphasized in industry. But these phases and activities are the basis for a process improvement of EC- and development processes because it can be identified whether the ECP was successful or not. Also the strategy learning depends on the late phases and a review of the EC in total and retrospectively. In conclusion, industry is currently abstaining from the potential to improve and increase the effectiveness and efficiency of the EC- and development processes by intensifying the late phases and extend the field of courses of actions.

In a next step methods and tools to assess change effects in order to support the decision making will be developed. Thereby, decisions within the phase identification for situation analysis purposes as well as decisions within the phase decision upon courses of actions are addressed. Furthermore, these methods and tools promise an improvement of the effort-benefit ratio, which is vital challenge in industry when estimating the effort to put in the elimination of target-deviations. A subsequent goal is to develop an approach for the strategy learning within ECPs and the evaluation of the benefit of this strategy.

Acknowledgement

We thank the German Research Foundation (Deutsche Forschungsgemeinschaft – DFG) for funding this project as part of the collaborative research center ‘Sonderforschungsbereich 768 – Managing cycles in innovation process – Integrated development of product- service-systems based on technical products’. Furthermore we would like to acknowledge the support of the participants of the industrial working group, who made a valuable contribution to this research.

11.6 References

Aßmann, G. (2000) Gestaltung von Änderungsprozessen in der Produktentwicklung, Diss., München, TU

Eckert, C., Clarkson, P. J., Zanker, W. (2004) 'Change and customisation in complex engineering domains', Research in Engineering Design, vol.15, no.1, pp. 1-21

Fricke, E., Gebhard, B., Negele, H., Igenbergs, E. (2000) 'Coping with changes: causes, findings, and strategies', Systems Engineering, vol.3, no.4, pp. 169-179

Huang, G. Q., Mak, K. L. (1999) 'Current practices of engineering change management in UK manufacturing industries', International Journal of Operations & Production Management, vol.19, no.1, pp. 21-37

Jarratt, T., Clarkson, J. (2005) 'Engineering change', in Clarkson, J. and Eckert, C. M. (eds.) (2005) Design process improvement, London, Springer, pp. 262-285

Jarratt, T. A. W., Eckert, C. M., Caldwell, N. H. M., Clarkson, P. J. (2011) 'Engineering change: an overview and perspective on the literature', Research in engineering design, vol.22, no.2, pp. 103-124

Kissel, M., Lindemann, U. (2013) 'System Architecture Change Decisions in Multi-variant Product Portfolios', International Conference on Engineering Design, ICED13. Seoul, Korea

Köhler, C. M. (2009) Technische Produktänderungen: Analyse und Beurteilung von Lösungsmöglichkeiten auf Basis einer Erweiterung des CPM/PDD-Ansatzes, Diss., Univ. Saarbrücken

Lindemann, U., Reichwald, R. (1998) Integriertes Änderungsmanagement, Berlin, Springer

Maier, A. and Langer, S. (2011) Engineering change management report 2011: Survey results on causes and effects, current practice, problems, and strategies in Denmark, Technical University of Denmark, DTU

Pahl, G., Beitz, W., Feldhusen, J., Grote, K.-H. (2007) Engineering design: a systematic approach, London, Springer

Rouibah, K., Caskey, K. R. (2003) 'Change management in concurrent engineering from a parameter perspective', Computers in Industry, vol.50, no.1, pp. 15-34

Rozenfeld, H., Forcellini, F. A., Amaral, D. C., de Toledo, J. C., da Silva, S. L., Alliprandini, D. H., Scalice, R. K. (2009) Gestão de desenvolvimento de produtos: uma referência para a melhoria do processo, São Paolo, Saraiva

Terwiesch, C., Loch, C. H. (1999) 'Managing the process of engineering change orders: the case of the climate control system in automobile development', Journal of Product Innovation Management, vol.16, no.2, pp. 160-172

Chapter 12

Consideration of Uncertainties in the Product Development Process

J. Reitmeier, T. Luft, K. Paetzold and S. Wartzack

12.1 Introduction

Modern product development processes are characterized by a high complexity and division of work. In addition, the cooperation of different engineering disciplines is needed to achieve the required product functionality. Therefore, product development is predominantly done in parallel and distributed processes. Disruptions in development projects that may result in a delayed development and time-consuming iteration cycles are the logical consequence if interfaces of individual work steps respectively processes are not sufficiently described or standardized. In addition, presumed minor changes to a single component often have an effect on a vast variety of overall system properties. As a result, the effect of wrong decisions and unnecessary iterations increases.

The recent years have shown that there is a strong trend to ensure product functionality by the means of simulations. This requires a pre-defined quality and processing of data and information depending on the current process step if virtual validations are to be executed efficiently. However, especially in early stages of a development process, lots of assumptions have to be made since data are not available or are subject to a considerable uncertainty. Thus, it is important to identify uncertainties, evaluate their effects and to formulate appropriate strategies to handle respectively reduce uncertainties.

The product's behaviour caused by the product's properties and by taking into account the specific usage and environmental conditions is a relevant measurement for the fulfilment of the desires and requirements. Therefore, it is the objective to develop a product-oriented process management based on a property-oriented product description. By using maturity and iteration management based on this and by the help of an appropriate simulation planning approach for virtual validation, it is not only possible to optimize the development process but also to improve the product itself. In this paper, the handling of uncertainties is focused and exemplarily shown by considerations concerning the design of a chassis.

M. Schabacker et al. (eds.), *Modelling and Management of Engineering Processes*, DOI 10.1007/978-3-662-44009-4_12

12.2 State of the Art and Preliminary Work

The almost unlimited number of ways in which functions desired by customers can be realized enforces a conceptual thinking according to Schäppi *et al.* (2005). Hence, the development and engineering work has been investigated scientifically by methodological and theoretical aspects since the middle of the last century (Vajna *et al.*, 2009). Pahl *et al.* (2007) define design methodology as specific procedures or instructions for developing and engineering of technical systems and products. The aim of these instructions is to enable a goal-oriented approach to facilitate the planning and team work and to serve as a guide for product developers (Pahl *et al.*, 2007).

12.2.1 The Property-based Product Development

Krehmer (2012) proposed an advanced procedure model in order to establish a product-oriented process management. This procedure model was further developed in Luft *et al.* (2013b and 2013d) and is composed of three main parts "procedure model", "micro-cycle of synthesis or analysis" and "matrix-based product description" and is shown in Figure 12.1 in a very simplified form.

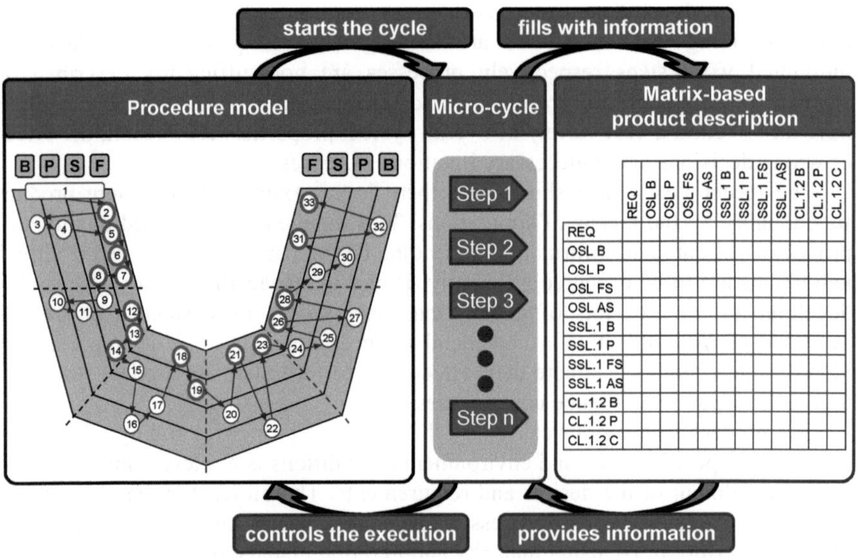

Figure 12.1. Simplified overview of the advanced procedure model according to Krehmer (2012) and Luft *et al.* (2013d)

The procedure model divides the entire product development process into 33 process steps which each of these is assigned to one of the four perspectives: behaviour (B), properties (P), structure (S) and function (F). It explicitly refers to process steps of synthesis and analysis (left and right part of the V-model) and therefore guides developers through the product development process (Luft *et al.*, 2013b). Based on the requirements (REQ), the product to be developed is described in detail top down regarding its overall system level (OSL), the associated subsystem levels (SSL) and component levels (CL). The system integration is progressively performed under continuous analysis of the achieved properties and behaviour at component, subsystem and overall system level. The micro-cycles assist in the processing of the process steps as they pretend what the developers have to do. Thereby, the matrix-based product description is step by step filled out with information regarding the respective behaviour, properties, characteristics (C) as well as the function structure (FC) and active structure (AS) of the OSL, the SSL and the CL. Thereby, similar definitions of characteristics and properties as defined in Weber (2005) are used. In turn, the matrix-based product description, which will be explained in the following section, provides information for the execution of the individual micro-cycles and thus controls the execution of the procedure model. Further explanations of the procedure model as well as of defintions are given in Krehmer (2012) and Luft *et al.* (2013b).

Due to the consequent detection and calculation of the product maturity, which consists of several key performance indicators (KPIs) (e.g. completeness of the collected requirements, configuration level of components; cf. Krehmer (2012)), at each step, the fulfilment of the required property profile can be monitored, and, as a consequence, need for action can be detected at an early stage, for example to avoid unnecessary iterations during the product development process (Krehmer *et al.*, 2010). By using the matrix-based product description, multiple dependencies, for example, of defined characteristics and resulting properties are mapped. Thus, deviations from targets together with their related causes can be recognized early. Therefore, better alternatives can be identified as well as their corresponding consequences can be estimated accurately. To sum up, this procedure model for iteration and product maturity management in the property-based product development supports virtual property validations and assists therefore developers during the product development process.

12.2.2 The Matrix-based Product Description

The main part of the procedure model is the matrix-based product description which consists of several Design Structure Matrices (DSM) and Design Mapping Matrices (DMM) and represents therefore a Multi Domain Mapping Matrix (MDM). By using such a matrix-based product description, all the dependencies and interactions within a product can be mapped during the development process systematically. By mapping the dependencies between characteristics and properties in a characteristic-property-matrix (see Figure 12.2), it can be analysed, which unintended effects on properties intended modifications of certain characteristics have (Luft *et al.*, 2013d). In addition, it is also possible to reconstruct which effects changes of certain characteristic have on the component, subsystem and overall system level. The matrix-based product description was evaluated by using the example of a chassis (e.g. wheel hub) and further developed (e.g. different types of dependencies) in Luft *et al.* (2013d). The interaction of dependencies between characteristics, properties and the resulting behaviour was shown by using the example of a front-wheel suspension. A simplified example matrix for the component wheel hub is shown in Figure 12.2.

Characteristic-(component)-property-matrix of a wheel hub		Properties				Characteristics					
		(A)	(B)	(C)	(D)	(a)	(b)	(c)	(d)	(e)	(f)
Properties	(A) Mass			1							
	(B) Bending stiffness				1						
	(C) Density	1									x
	(D) Elastic modulus		1								x
Characteristics	(a) Outside diameter						x	x		x	
	(b) Inside diameter					x					
	(c) Pitch circle diameter borehole					x			x	x	
	(d) Number of rim boreholes							x		x	
	(e) Pitch circle diameter brake disc					x		x	x		
	(f) Material			x	x						

Figure 12.2. Characteristic-(component)-property-matrix of a wheel hub (simplified extract)

A subsequent nominal-actual comparison provides information about whether desired properties of the product have been achieved satisfactorily or whether iterations are necessary (Krehmer, 2012). Consequently, it is important to analyse the flow of information by taking into account the involved employees as well as the single and distributed development steps (Luft *et al.*, 2013a). As a result, a property-based detection of the product maturity can be realized. Since the matrix-based product description is being developed mainly during the concept phase, not only the different uncertainties in this early stage of product development have to be considered but also the product's properties have to be validated.

12.2.3 Virtual Property Validation

A multitude of highly efficient simulation tools for different disciplines is available today (Paetzold and Reitmeier, 2010). However, it is still an open question which simulations can be executed at what point in time to really support product development processes and to reduce the development risk by purposeful and early validation of product functionality.

There are two fundamental kinds of process steps that are executed alternately: synthesis and analysis. Result of the synthesis are characteristics, while in the analysis step, the property profile is identified and compared against the requirements or specific reference values. The data processing of the network-like linkage of characteristics and properties via matrices, addressed in the previous section as well as in previous publications (e.g. Reitmeier and Paetzold, 2012), helps in the first step to identify necessary input data for specific property validations.

However, in the second step, it is absolutely necessary to indicate which input data the simulation results are based on in order to prevent systematic errors and avoid implying a precision that does not exist (Reitmeier and Paetzold, 2011): data quality must be pre-defined depending on the current process step if simulations are to be executed efficiently. The usefulness of a property validation is mainly determined by the quality of available data and information. DIN 55350 (1987) defines quality as the "entirety of characteristics (and values of characteristics) of a unit according to its ability to fulfil determined and preconditioned requirements". Based on the four quality dimension (intrinsic, contextual, representational and accessibility as crucial success factors) of Wang and Strong (1996), one of the most cited concepts when estimating information quality, quality attributes in the context of simulations are defined in Reitmeier and Paetzold (2011) and a correspondent evaluation system is presented.

In virtual product development, optimization cycles are defined based on simulation results, unless the realized property profile is not satisfying. In this context, the inclusion of sensitivity analysis cannot only support the basic filling (identification and weighting of linkages) of the matrix-based product description, but consequently also identify efficient “set-screws” (characteristics) to support the iteration management mentioned before. A correspondent scenario is presented in Reitmeier and Paetzold (2012) by the exemplary behavioural simulation of the break application of a car. In addition, results of sensitivity analyses can be used to identify the available overall data quality, as the quality of high important input data are rated higher than the quality of less important ones and, therefrom, the expected result quality of simulations. This supports to evaluate the usefulness and consequently the planning of simulations (Reitmeier and Paetzold, 2011).

However, the process design is strongly influenced by the boundary conditions of the development situation and the specific requirements for each development task and thus, methods and tools to purposefully support operational activities must be provided (Roelofsen, 2011). Ponn (2007) asserts that the development context can be described by context factors that have an influence on product and process. Hence, context factors are presented for an approach to simulation planning in Reitmeier and Paetzold (2013) and linked to the matrix-based product description. These factors include product-related (to identify the current development status: e.g. relevance of a property), process-related (to support the simulation planning and the evaluation of simulation results: e.g. availability or quality of necessary input data) and resource-related (simulation effort and basic feasibility of a simulation: e.g. comparison required/available capacity of tools or personnel) aspects to describe the development situation in the simulation context. Based on this, two essential aspects are considered within simulation planning (see Figure 12.3): the evaluation of the meaningfulness and feasibility of a simulation is put in front of a step of analysis and the subsequent evaluation of the validation results is supported by statements concerning the simulation quality.

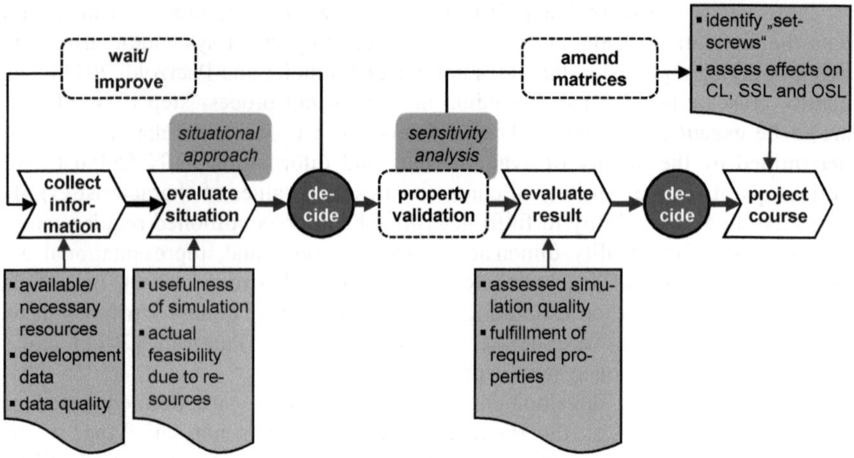

Figure 12.3. Basic concept of simulation planning following Reitmeier and Paetzold (2013)

12.3 Product-oriented Process Management

12.3.1 Consideration of Uncertainties in the Product Maturity Management

The matrix-based product description requires the use of information whose content particularly in early development stages is based partially on assumptions and therefore is highly uncertain. The level of uncertainty is compounded by the possibility that certain facts and figures may be completely unknown or disregarded (De Weck *et al.*, 2007). Therefore, it is necessary to develop a concept for the identification and handling of uncertainty in the development process to measure product maturity while taking into account the uncertainties. Thereby, the current product maturity should not be measured by the time spent and the costs incurred or in other words by time- and cost-related KPIs (cf. Luft *et al.*, 2013c) but instead of these by means of product-related KPIs (e.g. realized properties). For this, an appropriate approach for handling uncertainty in product development process is essential. Such an approach is depicted in Figure 12.4 and will be described briefly thereafter.

The starting point of the approach is the identification of uncertainties in a specific development project. Since development projects and therefore the respective uncertainties differ significantly from one another, a project-specific view is required. For example, uncertainties from numerous possible rim-tyre-combinations concerning a conventional car are different to these concerning a racing car where combinations are often limited by technical regulations. It should also be mentioned that (virtual) validations play a crucial role because the uncertainties are not in every stage of the development process from the same source and have the same degree. Therefore, it may be necessary to carry out the entire procedure for the analysis of uncertainties several times. Finally, the exact types of uncertainty for the identified uncertainties have to be determined. Following Derichs (1997), types of uncertainties can be classified into the two groups "context of uncertainty" and "content of uncertainty". Contextual factors of uncertainties are "changing frequency", "changing time", "changing amount" and "changing cause". The group "content of uncertainty" can be subdivided into three groups ("missing information", "incorrect information" and "misinterpretation of information") and is already integrated in the analysis of the development situation according to Reitmeier and Paetzold (2013).

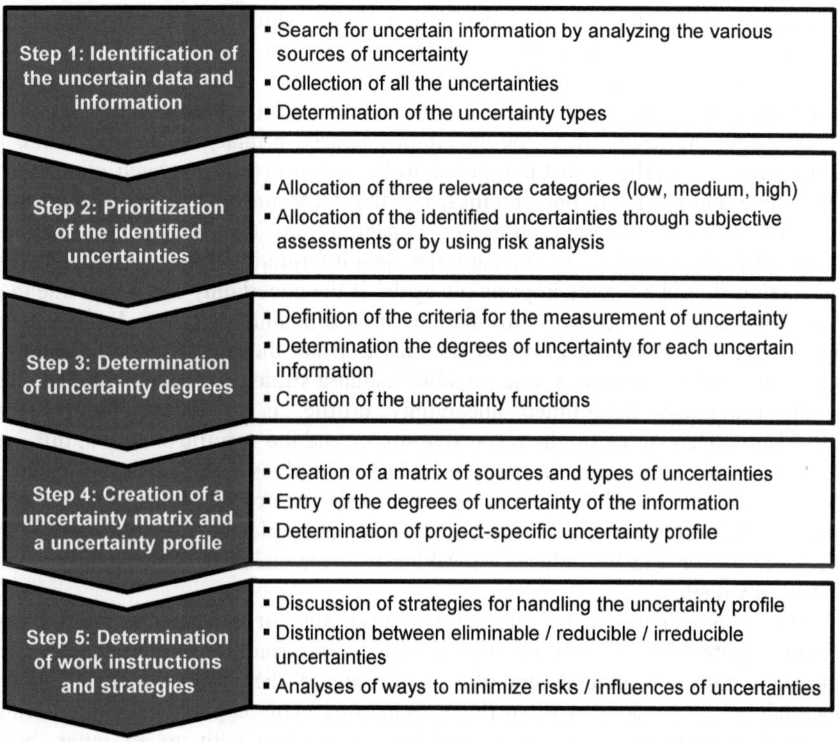

Figure 12.4. Overview of an approach for handling uncertainty in the development process

To keep the cost of the analysis of the uncertainties economically, the uncertainties have to be classified regarding their impact on product development in step two. For instance, it would be inefficient to analyse a certain uncertainty regarding the screw thread (e.g. M14 or M16 to fix the rim to the hub) to the last detail if it is ultimately irrelevant for the conceptual design phase which screws are used at all. This can be done through surveys of affected persons, using subjective assessments or through a risk analysis. The result of the second step should be a categorization of all identified uncertainties into three classes (low, medium, high).

After a successful prioritization of uncertainties, the effects have to be determined in the third step. All uncertainties are described by an uncertainty function. Subsequently, the degree of uncertainty of the analysed uncertain information can be determined from the uncertainty function which is a graphical representation of the relationship between the degree of uncertainty and the criterion of an uncertainty type. Since this procedure requires a relatively high effort, the focus should be on the uncertainties with high relevance according to the previous step. By identifying the different degrees of uncertainty, it is possible in principle to derive the measures which have to be performed to eliminate or at least to reduce these uncertainties.

This is followed in step four by the creation of the project-specific uncertainty profile by entering the previous results in a matrix which consists of uncertainty sources and types. This matrix is used to give an overview of the entire uncertainties of a development project and to show what causes most uncertainties of a project as well as what types and what degrees the uncertainties have.

Based on the results from the uncertainty matrix, primary fields of action can be identified. Thereby, a distinction has to be drawn mainly between eliminable, irreducible and reducible uncertainties. For the uncertainties, for which reduction potential is seen, different strategies depending on the source and type of uncertainty are possible, to increase the data/information security and quality. Distinction should be made in principle again, if the uncertainty has arisen due to a lack of information, incorrect information, or by a misinterpretation. For instance, for a lack of information, which can arise from the type's incompleteness, inaccuracy and doubtfulness, one possible standard strategy, which can be applied to the previously determined uncertainty profile, is to reduce the missing information. Depending on the respective source and the specific situation, this can be done by different work instructions. The doubtfulness (e.g. does the material "X" bear the stress "Y"?) regarding product properties or functions, for example, can be reduced by simulations (section 12.3.2). Incompleteness or inaccuracies of the CAD design can be reduced by asking the responsible engineer or by using historical values.

This approach has to be integrated in each step of the advanced procedure model (Figure 12.1), so that the influence of the uncertainties in the estimation of product maturity is evident and the monitoring of the product maturity under uncertainty can be done with sufficient reliability. This approach for monitoring the product maturity serves as a guideline for dealing with uncertainties, both during the product development process and in the assessment of the development progress. Furthermore, this approach supports to determine the usefulness of virtual validations due to the present (uncertain) data base and to evaluate their result quality. Due to the associated increase in the quality of information, wrong decisions especially in early development stages are reduced and the related occurrence of time-consuming and costly iterations is minimized. In addition, through the development accompanying product maturity monitoring (i.e. the comparison of actual and desired properties) in a variety of development projects, the strategic management of various development projects will be supported. This increases the overall efficiency in the product development of the company.

12.3.2 Management of Virtual Property Validations

As mentioned before, one core aspect of the research project is to trigger simulations efficiently and to substantiate decisions based on simulation results. In the context of simulation uncertainties occur mainly due to the quality of input data and simulation model and thus affect the accuracy and meaningfulness of results from this analysis. In addition, the effect of modifications of characteristics on properties (not only the original intended ones) is important to identify which properties have to be validated again. This is shown by considerations regarding the development of the chassis of a racing car with a view to process and product. In this context, the following main aspects are exemplarily illustrated:

- The simulation focus and related needs regarding data/information as well as the availability of data/information are based on the actual process step;
- The validation and optimization of component properties regarding the determined (possibly contradictory) component requirements;
- The effect of modifications (parameter changes of a single component as well as conceptual modifications) on component and system properties.

Core parameters of a chassis (e.g. pitch pole) show a dynamic behaviour depending on the lifting or steering of the wheel. Neglecting the elasticity of the components, the suspension is a defined kinematic chain that can be described by mathematical approaches (Heißing *et al.*, 2011). In doing so, a model with rigid body motion on trajectories is obtained that owns ideal articulations (ball joints without clearance are often used in race cars). Within chassis development, it is useful to determine the kinematic design with such a simplified model first and to expand and specify this in later stages of the development process (e.g. multi-body simulation with appropriate specialized tools to analyse elasto-kinematics and loads) (Heißing *et al.*, 2011). This shows that there are different requirements concerning input data and simulation model in different stages of the process. In addition, aspects like suspension geometry, material properties or boundary conditions like forces or moments are often based on assumptions in the early stages when there is not enough background concerning concepts, models or empirical values.

These statements point out that especially in early stages of a development simulation results are subject to uncertainty. This is based on the simulation focus (and corresponding level of abstraction of the model) as well as on the availability of data and information. This was initially discussed in Paetzold and Reitmeier (2010) and considered when determining context factors (Reitmeier and Paetzold, 2013) to provide targeted support for simulation planning and the evaluation of simulation results with respect to its quality that is depending on the quality of the used input data and simulation model.

The chassis of a formula student race car usually consists of five subsystems at SSL (wheel, suspension, spring-damper system, brake system, steering system). At CL, the wheel hub (see Figure 12.2) is an essential part of the suspension. Forwarding lateral forces from the wheel to the wheel bearing is an important function of it (Trzesniowski, 2010). Accordingly, the wheel hub must be designed for bending stiffness. Numerical simulations like stress and deformation analyses using finite element analysis are design standard (Heißing *et al.*, 2011).

But, in addition to appropriate strength requirements, the wheel hub must also meet the usual requirement for weight reduction in racing. Improvements in the vertical dynamics by reducing the mass of components may be exemplarily mentioned (Heißing *et al.*, 2011). For this purpose, star-shaped or ribbed (instead of massive) designs of the wheel hub are focused. In this context, the challenge is to achieve equal or higher cambering stiffness. If information regarding the production process (e.g. casting process) is available, geometry optimizations concerning an accurate production process can be considered as well.

These brief examples indicate the link-up of desirable properties and constructive designs. If a component property is not sufficiently fulfilled, optimization tasks have to be initiated. A matrix-based product description supports to identify efficient "set-screws" (characteristics). Furthermore, the effect of modifications on other properties is more transparent. This in turn supports on the one hand the initiation efficient iteration cycles and on the other hand to trigger other simulations when it is assumed that the iteration cycle may also have effects on other properties and, therefore, those need to be validated again.

In addition to the requirements on CL, aspects on SSL have to be considered as well. A wheel hub is used to fix the wheel. Here, a multi-screwed mounting (used for commercial cars) as well as a central screw mounting (preferred in racing with respect to a quick mounting) is possible (Heißing *et al.*, 2011). In the latter, the application of a screw thread on the hub is required. The outer end of the hub is not threaded. Here, an appropriate chamfer must be made to support a fast replacing of the wheel including a save guidance to the first thread. For example, formula student teams usually start with a multi-screwed mounting of the wheel and turn to a central screw mounting in a later generation of their car. The identification of effects of such a concept change on SSL and OSL properties can be supported by a matrix-based product description and point out important validations.

Another aspect on SSL to be noted is that wheel hub and wheel bearing must be harmonized. A wheel hub should be hollowly designed and with a diameter as large as possible, so that it is rigid and also light (Trzesniowski, 2010). However, the bearing friction is influenced by the size of the wheel bearing. Consequently, a constructive design has to consider properties on CL and SSL. Here, a matrix-based product description can trigger a corresponding analysis of the bearing friction when the diameter of the wheel hub is modified.

As well-known from racing, the choice of tire type has a great influence on the driving dynamic (requirement on OSL), also Heißing *et al.* (2011) emphasize that it is essential to consider all tire characteristics in this context. Therefore, when considering the driving dynamics of a conventional car, a multitude of tire sizes and models must be considered. In the racing, rims and tire combinations are often limited by technical regulations. This is another example that the matrix-based product description may not only support to show the effect of modifications on CL but also on SSL and OSL. Influencing parameters like tyre radius or profile concerning handling performance can be identified and quantified.

In conclusion, the previous statements show the networked link-up of possible constructional designs and required properties on CL, SSL and OSL. A product description byusing characteristics and properties can be supported by sensitivity analyses which the support the identification of appropriate "set-screws" for optimizations. In addition, it can be identified which data and information are required and important to execute specific property validations. The effects of assumed minor/major design changes (e.g. diameter of the wheel hub) can be captured more quickly and necessary property validations are triggered accordingly depending on the effect of modifications. Statements concerning their quality can additionally represent the validity of simulation results more transparent.

Finally, a simulation planning built on a matrix-based product description supports to better assess and consider uncertainties resulting from a specific point in time during the development process and from design changes (Figure 12.5).

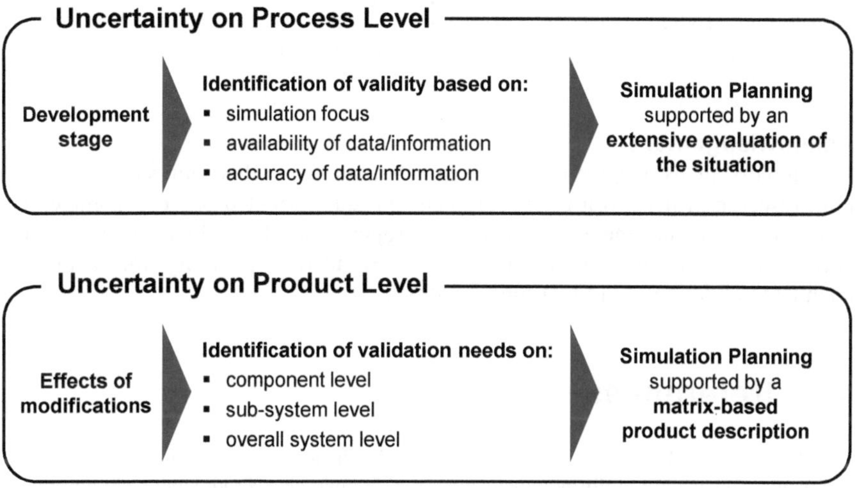

Figure 12.5. Handling of uncertainty by simulation planning

Besides a look at the resource-related basic feasibility and effort estimations, property validations respectively the validation needs are particularly initiated or rejected with respect to the range of modifications, uncertainty level of data and the boundary conditions of the process. This aspect is the more important, the more virtual development processes take place, as product maturity is estimated and iteration cycles are triggered based on simulation results. In this context, statements regarding the used data and models or the result quality of simulations can be considered as a measurement for uncertainty and a possibility to identify reasons for uncertainty.

12.4 Conclusion and Outlook

A product-oriented process management is introduced and illustrated using exemplary considerations when designing a chassis. In this paper, aspects to handle uncertainties that are based on process and product and influence virtual product development and property validations are focussed. The presented approaches to uncertainty management and simulation planning were developed in the research project concerning a product-oriented process management. However, these are also universally usable approaches.

Upcoming research focuses on a detailed cost-benefit analysis of the proposed procedure and a better integration of the proposed uncertainty management in the product-oriented process management. A first approach to this is to closer connect uncertainty evaluation and the evaluation of the usefulness of simulations in terms of the available data base. This not only supports to determine tasks of analysis with higher efficiency but also (especially in highly virtual product development processes) to improve the product maturity management.

Acknowledgement

The authors thank the German Research Foundation (DFG) for supporting this joint research project with the subprojects "Product-oriented process management – iterations management based on a property-based product maturity" and "Product-oriented process management – Simulation planning for a process attendant validation of product functionality".

12.5 References

Derichs, T. (1997) Informationsmanagement im Simultaneous Engineering - Systematische Nutzung unsicherer Informationen zur Verkürzung der Produktentwicklungszeiten. Aachen, Shaker Verlag

De Weck, O., Eckert, C., Clarkson, J. (2007) A classification of uncertainty for early product and system design. In: Proceedings of the 16th International Conference on Engineering Design, Paris, France

DIN 55350-11: 1987-05 (1987) Begriffe der Qualitätssicherung und Statistik; Grundbegriffe der Qualitätssicherung. Berlin, Beuth Verlag

Heißing, B., Ersoy, M., Gies, S. (2011) Fahrwerkhandbuch. Wiesbaden, Vieweg Teubner Verlag

Krehmer, H., Meerkamm, H., Wartzack, S. (2010) Avoidance of unnecessary design iterations by monitoring the product's degree of maturity. Proceedings of the 11th International Design Conference (DESIGN), Dubrovnik, Croatia

Krehmer, H. (2012) Vorgehensmodell zum Iterations- und Reifegradmanagement in der eigenschaftsbasierten Produktentwicklung. VDI-Verlag, Fortschrittsberichte VDI, Volume 1, Number 416

Luft, T., Bochmann, J., Wartzack, S. (2013a) Enhancing the flow of information in the PLM by using numerical DSMs – an industrial case study. In: Proceedings of the IFIP WG5.1 – 10th International Conference on Product Lifecycle Management, Nantes, France

Luft, T., Krehmer, H., Wartzack, S. (2013b) An advanced procedure model for property-based product development. In: Proceedings of the 19th International Conference on Engineering Design (ICED), Seoul, South Corea

Luft, T., Smoll, S., Wartzack, S. (2013c) An Indicator-Based Process Monitoring Cockpit for Controlling and Enhancing Product Development Processes – An Industrial Case Study. In: Proceedings of the 23rd CIRP Design Conference, Bochum, Germany

Luft, T., Wartzack, S. (2013d) Die matrixbasierte Produktbeschreibung als Bestandteil des Vorgehensmodells in der eigenschaftsbasierten Produktentwicklung. In: Proceedings of the Stuttgarter Symposium für Produktentwicklung, Fraunhofer-Verlag, Stuttgart

Paetzold, K., Reitmeier, J. (2010) Approaches for process attendant property validation of products. In: Proceedings of the 1st International Conference on Modelling and Management of Engineering Processes (MMEP), Cambridge, United Kingdom

Pahl, G., Beitz, W., Feldhusen, J., Grote, K.-H. (2007) Konstruktionslehre - Grundlagen erfolg-reicher Produktentwicklung - Methoden und Anwendung. 7. ed., Berlin, Springer

Ponn, J. (2007) Situative Unterstützung der methodischen Konzeptentwicklung technischer Produkte. München, Dr. Hut Verlag

Reitmeier, J., Paetzold, K. (2011) Evaluation of data quality in the context of continuous product validation throughout the development process. In: Proceedings of the 18th International Conference on Engineering Design (ICED), Copenhagen, Denmark

Reitmeier, J., Paetzold, K. (2012) Reduction of Uncertainty by Sensitivity Analysis and context-specific Data Processing within Virtual Property Validation. In: Proceedings of the 9th Norddesign, Aalborg, Denmark

Reitmeier, J., Paetzold, K. (2013) Process-integrated analysis of the development situation for an efficient simulation planning, In: Proceedings of the 19th International Conference on Engineering Design (ICED), Seoul, South Corea

Roelofsen, J. (2011) Situationsspezifische Planung von Produktentwicklungsprozessen. München, Dr. Hut Verlag

Schäppi, B., Andreasen, M., Kirchgeorg, M., Rademacher, F. J. (2005) Handbuch Produktent-wicklung. München, Carl Hanser Verlag

Trzesniowski, M. (2010) Rennwagentechnik. Wiesbaden, Vieweg Teubner Verlag

Vajna, S., Weber, C., Bley, H., Zeman, K., Hehenberger, P. (2009) CAx für Ingenieure. Eine praxisbezogene Einführung. 2. ed., Berlin, Springer

Wang, R. Y., Strong, D. M. (1996) Beyond Accuracy: What Data Quality means to Data Consumers. Journal of Management Information Systems, 12(4): p. 5-33

Weber, C. (2005) CPM/PDD - An extended theoretical approach to modeling products and product development processes. In: Proceedings of the 2nd German-Israeli Symposium on advances in methods and systems for development of product and processes. TU Berlin / Fraunhofer-Institut für Produktionsanlagen und Konstruktionstechnik (IPK), Stuttgart: Fraunhofer-IRB-Verlag, p.159-179

Chapter 13

Modelling Technique for Knowledge Management, Process Management and Method Application - A Formula Student Exploratory Study

A. Albers, N. Reiss, N. Bursac, L. Schwarz and R. Lüdcke

13.1 Introdruction

Due to increasing product complexity, an exchange of knowledge between different areas is getting more important in multidisciplinary product development. Furthermore, another trend can be recognized, which points out that employees change their position more often in an organization. Both trends, multidisciplinarity and a higher fluctuation lead to the consequence that it is necessary to explicate and share knowledge. Therefore, knowledge management should take up a central role in the product development processes (PDPs). Thus, a PDP has to be considered as a socio-technical system (Ropohl, 1975), which is characterized not only by technical development, but also is represented as knowledge-intensive process.

This paper reports on an ongoing research project IN² - from INformation to INnovation, which is funded by the German Ministry of Education and Research. It aims at research interactions between knowledge management, method application and PDPs. In order to optimize coupling of these aspects in a long term, an exploratory study in cooperation with the Formula Student Team KA-RaceIng is presented in this paper. It considers a modelling technique which allows the three above mentioned aspects to be modelled with the help of one technique.

The paper proceeds as follows. Section 2 gives a brief overview of the state of the art of method application, knowledge management and, especially, process modelling approaches. Section 3 focuses on the methodology and describes the framework of the exploratory study. In section 4, some findings of the exploratory study are looked at in detail. In Section 5, the conclusion and the outlook are presented.

M. Schabacker et al. (eds.), *Modelling and Management of Engineering Processes*, DOI 10.1007/978-3-662-44009-4_13

13.2 State of the Art

13.2.1 Method Application

In spite of large number of avaliable methods, only few of them are accepted (Wynn, 2010; Jänsch, 2007). Wynn criticizes a narrow focus and a high abstraction degree of method and process models, which contain too few recommendations about how they should be applied. Weiß determined particularly support deficits of the so-called "early stages" (Weiß, 2006). Tomiyama *et al.* have identified potential research fields based on the mentioned points of criticism (Tom, 2009). Referring to the investigated models and theories, a suggestion has been made to have a stronger focus in the scientific research on:

- complex continiously growing requirements;
- management of complex processes;
- stronger integration of different domains;
- ongoingly developing information and communication technologies;
- virtual engineering.

It can be determined that many through research developed methods are not used in practice.

13.2.2 Knowledge Management

Activities of product developers during PDPs can be described as ongoingly recurrent research, generation, validation and documentation of information. Knowledge emerges through interpretation, networking and detection of information (Klein, 2001; Willke, 1998; Polanyi, 1966; Wiig, 2008). The term information is often used synonymously to explicit knowledge; by this the knowledge objects are meant, which can be manipulated independently from the knowledge holder (files, documents, etc). In contrast to explicit, there is implicit and unspoken knowledge (Polanyi, 1966). NONAKA *et al.* have established a „SECI-Model" for description of differences and changes of knowledge upon explicating and implicating (Nonaka, 1995). This abbrevation stands for knowledge transformation: socialisation, externalisation, combination and internalization. Known methods and approaches for support of knowledge management address mainly such aspects as combination and internalisation (Abels, 2005; Reyes-Perez, 2009; Conrad, 2010). Knowledge management deals with generation, distribution and use of knowledge.

13.2.3 Process Modelling Approaches

There is a number of process modeling approaches in industrial practice. Thereby, the focus of each model is on different aspects. According to Wynn and Clarkson

(2005), a broad range of product development process models can be classified. They distinguish between stage- and activity-based models, between problem- and solution-oriented strategies, and abstract, analytical and procedural approaches. Flow charts and stage gate process models are suitable for the representation of straightforward procedures. Event-driven Process Chains (EPCs) embody a further approach for modeling business processes. The main idea of EPCs is that events release functions and, respectively, events are triggered by processed functions. EPCs were developed first of all for the purpose of process documentation. They can be distinguished by its intuitive and simple way of modeling as well as by high perceivability and good interpretability. In case of their extension to Business Process Model and Notation (BPMN), the responsibilities and distribution of tasks are added, what facilitates recording of organizational structures upon modeling. Currently known and established process models such as „methodology for developing and designing technical systems and products“ (VDI, 1993) as well as the V-Model (VDI, 2004) do not cope with modern trends. Mainly in process modeling a complete integration of method- and knowledge management into PDP is not sufficiently considered.

All the mentioned modeling approaches focus only on certain points, but do not consider an interaction between activities, knowledge objects, knowledge managements systems and methods.

Unlike the above mentioned approaches, the integrated Product engineering Model (iPeM) is an integrated approach which aims to fill in the gap between process management and engineering design. The iPeM shown in Figure 13.1 is a generic meta-model which comprises all relevant elements to derive situation-specific product development process models. Taking into account the real ongoing PDP, development of complex products takes place with the help of numerous, interconnected activities. These activities should be adapted to the changing boundary conditions (Meboldt, 2008; Albers, 2010). It is a framework for modelling flexible visualization of objects and their compounds in certain product design processes to facilitate an effective use of information and to allow recording of knowledge. The activities of the iPeM are divided into macro and micro activities (Albers, 2010). Micro activities appear iteratively in technical problem solving, whereas macro activities provide areas of product engineering (Albers, 2011). The iPeM is based on the system triple of product engineering: it describes product engineering as a continuous interaction of the system of objectives, the system of objects and the operation system (Ropohl, 1975; Braun, 2013) The system of objectives contains all explicit objectives (i.e. goals and their constraints) as well as their interrelations (e.g. conflicts) and justifications ("design rationale"). The system of objects contains all synthesized artefacts (knowledge objects). The operation system is a socio-technical system that contains structured activities, methods and processes. Additionally, it comprises involved roles and required resources.

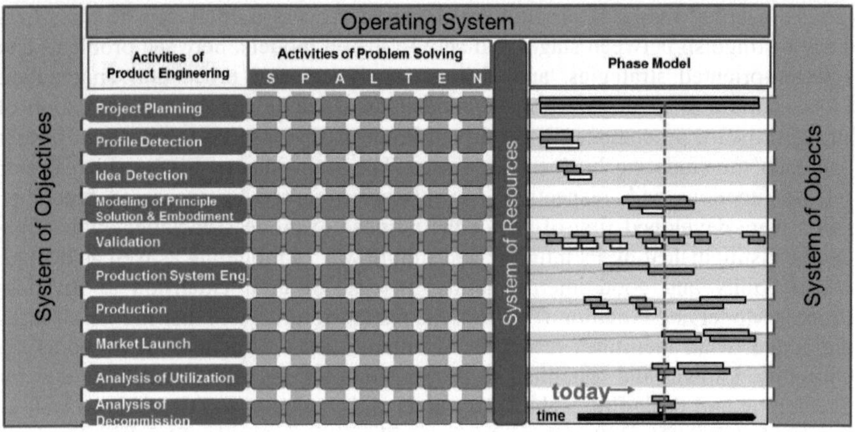

Figure 13.1. iPeM - Integrated Product Engineering Model (Braun, 2013)

Since the iPeM is a meta model, with which no real PDPs can be directly created, there is a need for a pragmatic applicable modelling technique which addresses these central ideas. With this technique, it is possible to model activities, methods, tools and further resources such as knowledge flows, knowledge objects and Knowledge Management Systems (KMS).

13.3 Methodology

Considering the state of the art, it can be determined that PDPs are insufficiently supported by the targeted method application and knowledge management. All the mentioned modeling approaches focus on individual aspects, but do not consider an interaction between activities, knowledge management and method application. Therefore, a technique will be presented, which allows a consideration of the three aspects and, subsequently, will be applied to an exploratory study, at the Formula Student Team KA-RaceIng in order to identify optimization potentials.

13.3.1 Methodolody and Structure of the Paper

Figure 13.2 visualizes the structure and methodology of the research project IN [2]. A theoretical framework of the current state of the art with respect to process models and their modelling served as basis for the research project. Based on this, a modelling technique is presented, which allows a representation of processes, knowledge and method management in one model. With the aid of the developed modelling technique, the PDP of the Formula Student Team KA-RaceIng will be shown as part of the explorative study. Thus, based on the obtained from the Formula Student Team process map, optimisation potentials were derived in the next step, and recommendations for further actions are stated. This contributes to the validation of the modelling technique.

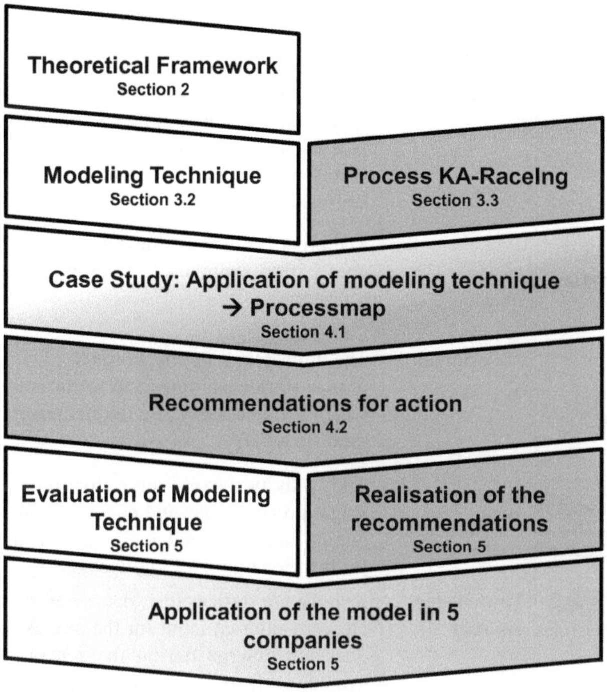

Figure 13.2. Methodology and structure of the paper

Based on the findings of the exploratory study, in a following paper the processes of 5 consortium partners of the IN^2 project will be analysed with regard to the targeted method and knowledge management.

13.3.2 Modelling Technique

For the purpose of connecting process- and knowledge management, an activity based diagram was developed based on the underlying concept of the iPeM. An activity is defined as a transformation of one knowledge-object into another. It can be supported by tools and methods. The resulting models provide information about the use of methods and the state of the art of Knowledge-Management-Systems in the considered companies. The term "knowledge object" can be defined as explicit knowledge, which can be manipulated independently of the knowledge holders (e.g.: files, documents, drafts, etc.). In addition, methods were added to the diagram, which support various activities in the product development processes.

In Table 13.1 the symbolism for the modeling technique is outlined.

Table 13.1. Elements of the modelling technique for the representation of knowledge management, method application and PDPs

Elements	Name	Explenation
	Knowledge holders	Knowledge holders are people, who "carry" certain knowledge objects. Their classification is specified in horizontal swimlanes, which are labeled in the left column. Processes, which are represented in the swimlanes, are carried out by knowledge holders.
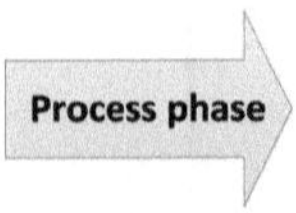	Process phase	The process phase is illustrated in a horizontal row above the swimlane in form of arrows and milestones.
	Milestones	The milestones define borders of single process phases from each other and set intermediate goals, which are necessary for the management point of view on a PDP.
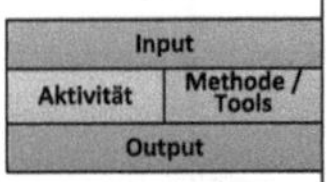	Activity Boxes	The activity boxes serve for representation of single process steps and consist of an input and an output, activities and methods that are necessary for the preparation of knowledge.
	Knowledge sources	Knowledge sources (e.g. datasheet, material tables, etc.) are external input for the processes. The input does not emerge during the process phase. It is available from the beginning and, thus, is especially important in the concept phase.
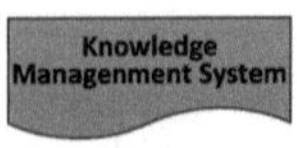	Knowledge Management Systems	Knowledge Management Systems (KMS) are the processes' knowledge archives. Generated and researched data is stored in them, so that it is accessible at certain time in PDP.
	Knowledge flows	Arrows connect the activity boxes and make up knowledge flows, which flow from one process step to the next one.

13.3.3 KA-RaceIng Process

The validation of the process model is realised by monitoring an engine development, carried out by the Formula Student Team KA-RaceIng in cooperation with an industrial partner. The formula student team was chosen because of its high level of interdisciplinary (mechanical-, electrical engineers, computer scientists e.g.) on the one hand, and a high turnover (annual change of composition) on the other. KA-RaceIng develops designs and builds two Formula Student racing cars, one combustion car and one with an electrical drivetrain, per Formula Student season. This means that each season a team of about 65 students, which are mainly new team members, has one year to build two racing cars in order to compete in the international Formula Student events with other teams. Due to the short development time and the high fluctuation, this project can only be successful with a well organised and carried out knowledge management. All generated information and all process stages have to be documented precisely, considering each component of the cars. Moreover, for team structure like the one of KA-RaceIng, meaning a high number of people, who have been a part of the project in the past, and still are available for passing on the information, it is very important to be able to reproduce when information was generated and who has generated it.

13.4 Findings

13.4.1 Process Map

Throughout six months the process was documented in cooperation with the project management of the engine development with a focus on knowledge management. Interviews and consultation supported the creation of the model throughout the entire modelling process (Figure13.3).

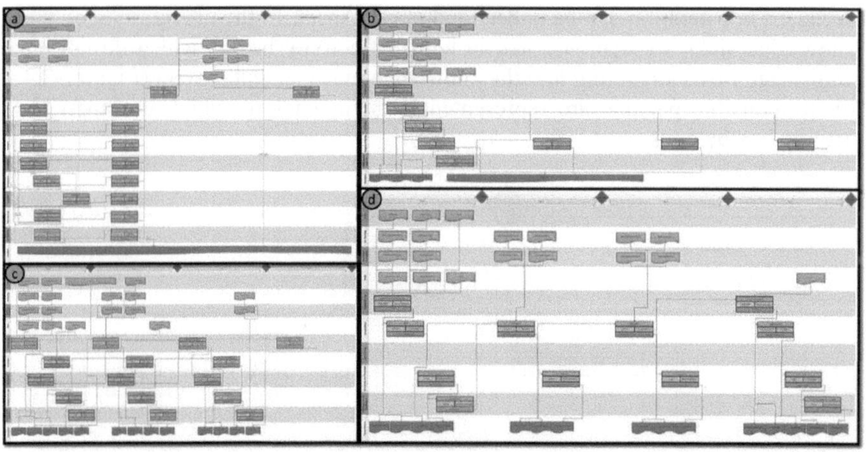

Figure 13.3. Process map of: a) overall process, b-d) sub processes

The objective was to generate a documentation in order to ensure traceability, despite the high fluctuation. Moreover, weak points of the development process and the applied knowledge management were revealed, resulting in a demonstration of optimisation potential. The first step was a modelling of the engine product development process and a collection of the generated knowledge objects, as well as, the correlation of these objects with knowledge holders and the degree of maturity. This leads to placement of the knowledge in the process followed by the analysis of activities, methods and tools used during the individual process steps in order to transform knowledge objects in the so-called activity boxes. The model emerges from insertion of the activity boxes into the process map consisting of the course of the process, knowledge holders, information sources and data bases. Interviews with the respective knowledge holders result in a network of knowledge objects via knowledge flows.

13.4.2 Recommendations for Action

Based on the process model, different optimization potentials could be identified during the KA-RaceIng exploratory study.

13.4.2.1 Identification of knowledge sinks

By reference to the modeled knowledge flows occurring during the PDP, knowledge sinks can be identified. Knowledge sinks may result from the insufficient knowledge management of a certain process stage. There are two different cases of knowledge sinks which cause different problems. Firstly, they occur if information is passed on to a person that does not need and use it and, therefore, does not pass it on further (Figure 13.4 – a). This may result in a waste of resources and capacity. Knowledge flows, which were not documented in the WMS in the course of several activities represent another problem. Neither decisions nor iteration can be understood in retrospect. In the process model, these knowledge sinks are demonstrated in terms of activity boxes, which do not have knowledge flows streaming into the knowledge management system (Figure 13.4 – b). Responsible persons are represented by swimlanes to which the approprate activity box belongs.

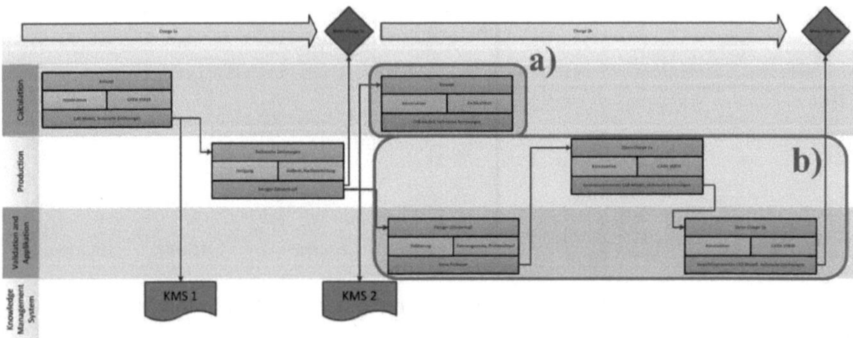

Figure 13.4. Example for the Identification of knowledge sinks

As a recommendation for action, these knowledge sinks should be pointed out, so that a knowledge loss can be avoided if the concerned role leaves the project.

13.4.2.2 Missing documentation concerning certain process activities

It becomes obvious, regarding the process model of the exploratory study, that certain process activities are documented less than others. It can be seen that certain departments tend to simply pass on information via communication instead of documenting it in databases. This non-documented knowledge can easily get lost and not used in the next process stage. Besides, it can lead to a repetition of errors due to missing information. In the model this can be seen by the lack of knowledge flow arrows from the validation into the databases. In the process model this is demonstrated by swimlanes that only have knowledge flows to other swimlanes but not to databases. In such cases, the reasons for not documenting the knowledge should be investigated.

13.4.2.3 Control of the process-accompanying documentation

The process model of the considered exploratory study indicates the decreasing quality of documentation and knowledge management as the process progresses. An effective knowledge management may be characterized by a steady documentation of the generated information. This is demonstrated in the process model by arrows of knowledge flow from each activity box to the corresponding database. The process model of the engine development demonstrates that with progressing process there are fewer arrows and, therefore, a decreasing knowledge flow. The modeling technique enables control of the process-accompanying knowledge management throughout the whole process. Moreover, it serves as guidance for following processes and projects.

13.4.2.4 Milestones as control points

At the end of a process stage indicated by a milestone, there should be a control of the determined information of the certain PDP in order to guarantee a process-accompanying documentation throughout the PDP. Based on the process model, it can be seen which knowledge objects are needed in following process steps and where are they stored when process stage comes to a milestone. Moreover, a final report, of what has been done during this process stage and how it was done, may have advantages for the further development. Especially, the lessons learned during this process stage need to be documented in order to prevent the repetition of errors.

13.4.2.5 Use of different knowledge management systems

There is a great variance in practice when it comes to the use of the different knowledge management systems (KMS). This can be seen in the model in Figure 13.5: there are knowledge flows in more than one KMS after each activity. For example, important dates are stored in the wiki (KMS 1) and other databases (KMS 2). This may cause a lack of information for persons who only have access to the wiki.

As a recommendation for action, the access rights of the KMS should be annalysed to figure out which ones can be combined in order to reduce the number of knowledge management systems.

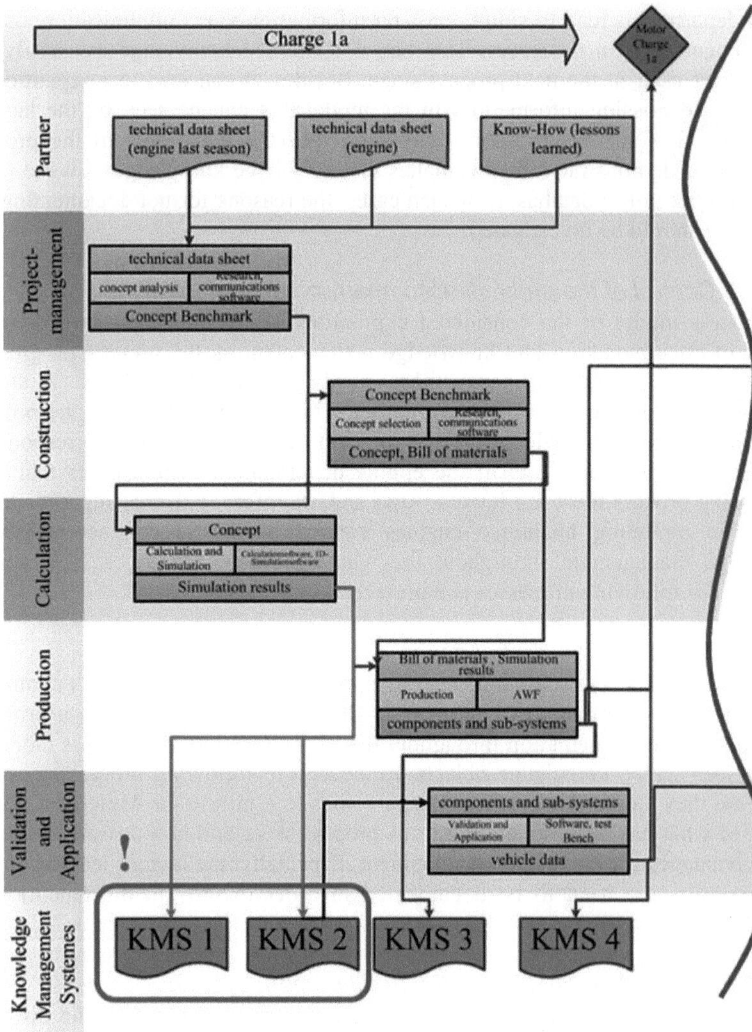

Figure 13.5. Use of different knowledge management systems

13.5 Conclusion and Outlook

The Formula Student Team was observed during six months, and various recommendations for actions were identified with the help of the new modelling technique, which allows a consideration of the following aspects: method application, knowledge management and process management. Thereby, it was shown that with the help of the new modelling technique optimisation potentials in an organisation can be pointed out. Hence, it seems to be useful to investigate the interactions between these three aspects.

It remains to be examined, which of the generated recommendations can be implemented in the PDP of the Formula Student Team KA-RaceIng, and how much impact on effectiveness and efficiency it has. Based on the process model, there are further steps recommended in order to improve the knowledge management and, consequently, the PDP of the engine development by the Formula Student Teams. These recommendations have been carried out in the considered process. An important step will be the introduction of a control of the different departments' knowledge management, carried out by the announced responsible person or department. This will be realized by control points at every milestone, at which final reports need to be written by each department. At these control points, in parallel to checking the final reports, it will be controlled if every important action and decision have been documented in the according database. Furthermore, it will be taken into consideration that a redundancy in the documentation might rather be confusing than helpful. Another step will be an identification of the departments that do not document the generated information. After this identification the reasons for documents not being documented and the possibilities of changing the circumstances, that enhance the lack of documentation, will be analysed. At last, based on the process models, it is possible to give people, working in the process, an understanding of knowledge management and its significance in PDPs. In a following work the Formula Student Team should be observed again in order to identify how the recommendations for actions will be implemented, and which influence will they have on the course of the project.

Furthermore, in an upcoming work, the modelling technique will be applied at five industry partners from different industries ranging from 40 to 40.000 employees, to analyse their PDPs and to figure out if the same recommendation for action can be made.

Acknowledgments
This research and development project is funded by the German Federal Ministry of Education and Research (BMBF) within the Framework Concept "Research for Tomorrow's Production" and managed by the Project Management Agency Karlsruhe (PTKA).

13.7 References

Abels, S., Haak, L., Hahn, A. (2005) "Towards Semantic Model Integration for Knowledge Management Systems." In: *Proceedings of the 3rd Conference Professional Knowledge Management - Experiences and Visions. Proceedings der GI, Kaiserslautern, Althoff, K. D. Dengel, A., Bergmann, R.; Nick, M.; Roth-Berghofer, T.*

Albers, A. (2010) "Five Hypotheses about Engineering Processes and their Consequences", In: *Tools and Methods of Competitive Engineering TMCE 2010, Ancona, Italien*

Albers, A., Braun, A. (2011) "A generalised framework to compass and to support complex product engineering processes". In: *International Journal of Product Development 15 (1/2/3), pp. 6–25*

Braun, A., Ebel, B., Albers, A. (2013) "Activity-Based Modeling and Analysis of Product Engineering Processes". In: *Smart Product Engineering: Proceedings of the 23rd CIRP Design Conference, ed. by Abramovici, M., Stark, R.*

Conrad, J. (2010) Semantische Netze zur Erfassung und Verarbeitung von Informationen und Wissen in der Produktentwicklung, Dissertation, Universität des Saarlandes

Horváth, P., Möller, K. (2004) Intangibles in der Unternehmenssteuerung. Vahlen Verlag, München

Jänsch, J. (2007) Akzeptanz und Anwendung von Konstruktionsmethoden im industriellen Einsatz. Analyse und Empfehlungen aus kognitionswissenschaftlicher Sicht

Klein, G. (2001) Wissensgesellschaft - Ein Schlagwort aus ungewöhnlicher Perspektive beleuchtet. In: *Graf, H.G. (Hrsg.): Und in Zukunft die Wissensgesellschaft. Der Umgang mit Wissen im Entscheidungsprozess. Verlag Rüegger, Zürich, pp. 81-88*

Meboldt, M. (2008). Mentale und formale Modellbildung in der Produktentstehung-als Beitrag zum integrierten Produktenstehung-Modell (iPeM). Forschungsberichte / IPEK, 29. IPEK Dissertation, Universität Karlsruhe, Institut für Produktentwicklung, Karlsruhe

Nonaka, I., Takeuchi, H. (1995) The Knowledge-Creating Company - How Japanese Companies Create the Dynamics of Innovation. Oxford University Press, New York

North, K., Güldenberg, S. (2008) Produktive Wissensarbeit(er): Antworten auf die Management-Herausforderung des 21. Jahrhunderts. Gabler Verlag, Wiesbaden

Polanyi, M. (1966) The Tacit Dimension, New York: Doubleday

Reyes-Perez, M., Gausemeier, J., Nordsiek D. (2009) Ontology Development for a Manufacturing Data Base for Products with Graded Properties. In: *International Conference on Information, Process, and Knowledge Management, pp. 105-109*

Remus, U. (2002) „Prozeßorientiertes Wissensmanagement - Konzepte und Modellierung" Universität Regensburg, Dissertation

Ropohl, G. (1975) „Einleitung in die Systemtechnik", In Systemtechnik – Grundlagen und Anwendung, München Wien, Carl Hanser Verlag, pp. 1–77

Tomiyama, P., Gu, Y., Jin, D., Lutters, C., Kimura, F. (2009) Design methodologies: Industrial and educational applications, CIRP Annals - Manufacturing Technology

Verein Deutscher Ingenieure (1993) "VDI-Richtlinie 2221, Systematic approach to the development and design of technical systems and products", VDI-Verlag, Düsseldorf

Verein Deutscher Ingenieure (2004) „VDI-Richtlinie 2206, Design methodology for mechatronic systems", VDI-Verlag, Düsseldorf

Weiß, S. (2006) Konzept und Umsetzung eines Navigators für Wissen in der Produktentwicklung. Fortschritt-Berichte VDI. 386.

WIIG, K. (2008) Knowledge Management for the Competent Enterprise. Business Intelligence Vol. 8, No. 10. Cutter Consortium

Willke, H. (1998) Systemisches Wissensmanagement. Lucius & Lucius, Stuttgart

Wynn, D. C., Clarkson, P. J. (2005) Models of designing, In: *Design process improvement – A Review of Current Practice. Clarkson P. J., Eckert C. M. (Eds.), Springer, London*

Wynn, D. C. (2010) "How Can PD Process Modelling Be Made More Useful? An Exploration of Factors Which Influence Modelling Utility" In: *International Design Conference DESIGN 2010, Dubrovnik*

Chapter 14

Alliance Management Process Design with Failure Mode and Effect Analysis

B. Tuna and H. Behret

14.1 Introduction

In today's competitive environment, developing mutually beneficial alliance has become inevitable for organizations to ensure sustainability and to achieve their strategies. That's why finding the most appropriate candidates for alliance, selection, development and management of alliance is important as well as identifying the issues on how to be a cooperate and to which purposes to be served.

The design of the alliance management as an administrative process has great significance for organization strategies. In the literature, alliance management is mostly examined in terms of strategic or as a supplier selection process. However social and operational dimensions are rarely considered in this process which we should define all together to make multi-dimensional analysis indeed. It is obvious for global companies that they should consider their partnership in social and operational dimensions to maintain sustainability and support their strategies through operational excellence. For this reason, we propose an integrated administrative model combining strategic, social and operational dimensions for establishing alliances (partnerships) which also bring a new insight for the current literature.

Well organized processes help companies to do their operations easier and get the cost advantage by taking into consideration the customer satisfaction. For this reason, alliance management processes should be designed and managed accurately considering process management principles and companies strategies. In this study, Failure Mode and Effect Analysis (FMEA) is used in order to determine the main process flow of alliance management and to find the process design criteria considering potential failure modes through this process. FMEA is a systematic set of activities intended to help a designer or engineer to analyze the design of a system (product or process). Based on the core business process steps and possible potential problems, FMEA helps to determine how and with which machines, equipments and issues can the process run accurately. Here, the main objective is to identify the main process flow and the sub processes of alliance

M. Schabacker et al. (eds.), *Modelling and Management of Engineering Processes*, DOI 10.1007/978-3-662-44009-4_14

management systematically by FMEA. The results of the analysis will identify the most important steps of the process design, risks and failures of the system. Thus, a precise alliance management process design can be maintained.

14.2 Methodology

14.2.1 Alliance Management

In today's world, the capabilities of the enterprises to achieve numerous goals are becoming inadequate and they require cooperations (Bamford *et al.*, 2003). For this reason, today's organizations developing mutually beneficial cooperations in order to ensure sustainability and to develop strategies became inevitable. Therefore, besides determining the necessities such as how cooperations should be or to which purposes would it serve, some main issues such as finding and choosing the most appropriate partner candidate which appropriately suits company's strategies, managing and developing successful cooperations are also particularly important.

The general definition of cooperation is the process of groups of organisms working or acting together for their common/mutual benefit. An alliance is a pact, coalition or friendship between two or more parties, made in order to advance common goals and to secure common interests.

According to Association of Strategic Alliance Professionals (A.S.A.P., 2002), the definition of an alliance is a close, collaborative relationship between two or more entities that share complementary assets and strengths to create increased value for their customers and their own organizations that could not be accomplished independently.

According to many researchers, alliances can be defined as a voluntary, evolutionary and flexible organization forms (Osborn and Hagendoorn, 1997), between two or more organizations (Duysters, 2001), to realize both collective and individual goals (Varadarajan and Cunningham, 1995).

Alliance management enables companies to provide high quality products and services with lower costs, to reach the resources such as expertice, technology and raw materials easily and to direct a wider geography or new segments of an established market. When we examine the literature about alliance management, we noticed that, alliance management is mostly examined in terms of strategic or as a supplier selection process. However, global companies should also consider their cooperations in social and operational dimensions in order to maintain sustainability and support their strategies through operational excellence. In this study, we propose an integrated administrative model considering strategic, social and operational dimensions of alliance management.

14.2.2 BPM and Process Design

In today's competitive environment, companies must design and manage their process in the lean manner based on the value chain in order to survive and to gain advantage over the competitors. By operating in a well-designed processes structure, the companies can have more customer satisfaction with the lower costs and with simple and standardized operation steps. Davenport and Short (1990) emphasizes that the requirements for business process design are diligence and creativity. Out of company borders the alliance with suppliers, customers, society, stakeholders has great significance on achieving the business goals of the company. Therefore, the alliance management process must be well-designed by relying on process management principles to implement our strategies.

Aras (2005) states that process management is the management of the processes in a systematic way. Process management aims to provide the definitions of all the activities in the process manner.

Business Process Management (BPM) is based on the necessity of the value added tasks to be performed in the administrative processes. BPM focuses on the management of business processes in an efficient and effective way (Jeston and Nelis, 2006). BPM is a management approach to achieve efficiency by integrating the business process and to improve the process in accordance with the corporate strategies, (Alexieva, 2012). Davenport and Short (1990) defines the business process as a set of logically related tasks in order to perform the desired business outcome.

Jeston and Nelis (2006) also mention that BPM can be considered as a two-way structure management of business processes as an integral part of management and management of business process improvement.

American Productivity and Quality Center (APQC) classify the process into three groups: operational, administrative, and support processes. The operational processes include all the processes between perceiving the customers' needs and expectations and offering service to customers. On the other hand, administrative and support processes include operational steps related with the management of human resources, information technologies, external relations and development.

In the study of Jeston and Nelis (2006) it is indicated that the processes are basically grouped in three levels:

- Strategic processes – this level represents the strategic processes, which must ensure that the underlying processes are meeting and /or continue to meet the specified objectives.
- Core processes – this level represents the core, or main business activities of the organization.
- Support processes – this level represents the non-core processes, which support the core processes of the organization.

Operational processes directly focus on the formation of the products and services perceived by the customer. However, administrative and support processes focus on the quality, management and services.

In this study, alliance management is considered as a business process. Hence, the process design is performed based on business process management (BPM) principles. In the first step of BPM, the requirements of the company should be taken into consideration while designing a process. In a general approach the phases of the process design are preparation, defining the objectives, items, owner, and relations with other processes, determining the performance indicators, pilot implementation and documentation of the process (Aras, 2005).

In the content of our study the phases of the process design for alliance management which is considered as an administrative process is determined as following: (1) The occurence of the need for the alliance; (2) Decision of the alliance formation according to the needs; (3) Building the inventory structure for the potential partners to select the feasible alliance to meet the needs of the company; (4) Managing the inventory structure in terms of the current and potential partners for strategic alliance; (5) Selection of the partners among the alternatives; (6) Managing the alliance; (7) Conclusion of the alliance; (8) Performance evaluation of the alliance; (9) Risk evaluation of the alliance and finally; (10) Inventory updating and final documentation of the alliance. The basic steps of the alliance management process is shown in the Figure 14.1.

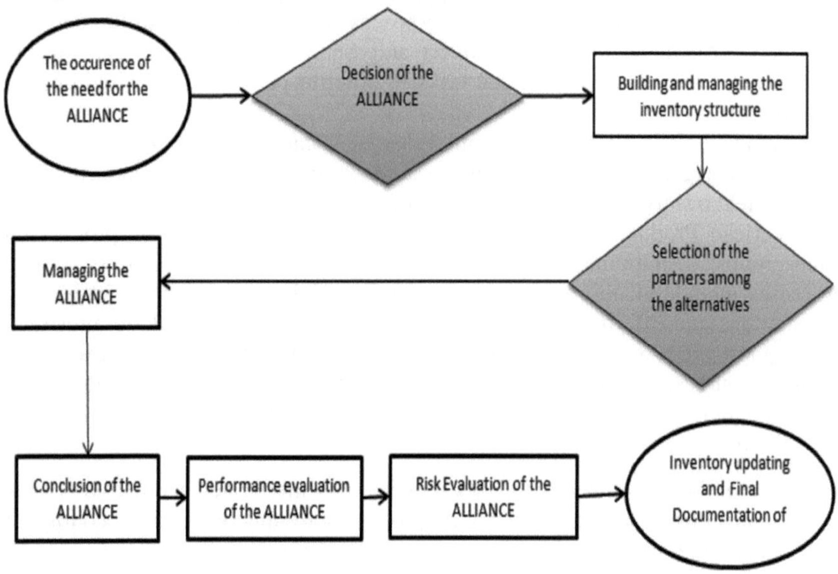

Figure 14.1. Alliance Management Process

This workflow determines the scope of our study as well. The analysis of FMEA will cover all the steps defined above in the process workflow. All the requirements and the actions for each steps will be mentiond in the analysis.

14.2.3. Process Design with FMEA

FMEA is systematic approach to identify and prioritize potential failures before they occur (Chen and Ko, 2009). In the literature there are a large number of studies related to the FMEA.

FMEA technique was first reported in the 1920s and it was used by National Aeronautics Space Agency in order to improve the reliability of military equipment in the 1960s. (Johnson and Khan, 2003).

Sellappan and Palanikumar (2013), the traditional FMEA uses Risk Priority Number (RPN) to evaluate the risk level of a component or product/process which is determined by the severity, occurrence and detection indexes (Table 14.1). RPN is an indicator of the risk level of the part's failure mode in design stage, determined by the multiplication of three characteristics, the severity of the potential failure (S), the frequency of potential failure (O), and the detectability index (D), (Chen and Ko, 2009).

$$RPN = S \times O \times D \qquad (14.1)$$

Table 14.1. S-O-D Effect Rating Scale for FMEA (Sellappan and Palanikumar, 2013)

Rank	Severity (S)	Occurrence (O)	Detection (D)
10	Hazardous without warning	Extremely high	Absolutely uncertain
9	Hazardous with warning	Very high	Very remote
8	Very high	High	Remote
7	High	Frequent	Very low
6	Moderate	Moderate	Low
5	Low	Occasion	Moderate
4	Very low	Slight chance	Moderately high
3	Minor	Very slight chance	High
2	Very minor	Remote, very unlikely	Very high
1	None	Extremely remote	Almost certain

Based on the descriptions in the literature, FMEA can be defined as the technique used to define the failure modes and to take preventions in the design, improvements and the operations of the processes. In FMEA, RPN is obtained by determining the severity, occurrence and detection indexes in the processes and products. Obtained the RPN scores, we examine the failure modes beginning from the highest score and determine the measures against these failure modes. Highest RPN sores indicate that we encounter the related failure mode unless we take necessary measures. In our study, initially, the potential failure modes for each process steps are determined and then RPN scores are calculated. So that we can start implementation studies based on the predetermined RPN scores.

In the study of Chen and Ko (2009) a conventional form of FMEA includes the following items:

- the design function of parts/process,
- the potential failure mode (categories of failure),
- the potential effects of failure (measured by the severity index),
- the potential causes of failure (measured by the occurrence (frequency) index),
- the detection method (measured by the detectability index)
- the risk priority number (RPN).

Process FMEA is a special kind of FMEA which is used to make process design considering the desired content. Process FMEA is a collection of possible causes and mechanisms for failure modes based on knowledge and experience used to develop a process. The purpose of FMEA method is to identify the weaknesses of the process and then to take measures for them. In the first stage of the analysis, the basic process steps are indicated. Then possible failure modes for each process steps are identified. Possible failure modes can have alternative causes for each process steps. These defined failure modes considered to have bad impacts on the process are scored in terms of severity, occurrence and detectability to calculate a parameter Risk Priority Number (RPN) which is the focus of our analysis.

In the concept of this study process FMEA is used for alliance management process design. Firstly, potential failure modes, potential effects of failure, potential causes of failure, current process control and recommended actions are determined based on the defined steps of alliance management process. Afterwards, the severity, occurrence and detection scores are determined and RPN scores are evaluated. Before the evaluation of each index the team should agree on the evaluation criteria for Severity, Occurrence and Detection Rating scales. The rating scales for the Severity, Occurrence and Detection are specifically adapted and modified in order to meet our project needs, thus causing several differences from the literature.

Severity (S): Severity is the score associated with the most serious effect for a given failure mode. Severity is the score that indicates the level of impact for stakeholders caused by potential failure modes in each process steps. The Severity score calculation table is shown in Table 14.2. The scores are defined as, "10 – High Importance" and "1- Slight Importance". All Possible failure modes are scored between 1 and 10 based on their level of impact.

Table 14.2. Severity Rating Scale for FMEA

Score	Criteria*	Modified Criteria
10	Hazardous Effect	May cause damage to stakeholders
9	Potential hazardous effect	Classified as illegal
8	Customer very dissatisfied	Makes the product or service is not used
7	Customer dissatisfied.	Creates a large stakeholder dissatisfaction
6	Customer experiences discomfort.	May Cause partial breakdown
5	Customer experiences some dissatisfaction	May result in significant loss of performance
4	Customer experiences minor nuisance	May result in slight loss of performance
3	Customer slightly annoyed.	Effective but loss of effect can be avoided
2	Customer not annoyed	Undetectable and slight effect on performance
1	No effect.	Undetectable and no effect on performance

* *Narayanagounder and Gurusami (2009)*

Occurrence (O): Occurrence is the probability that a specific cause will occur resulting in the failure mode within the design life. Occurrence is the score indicating the frequency of the potential cause and possible failure mode. The Occurrence Rating scale is indicated in the Table 14.3.

Table 14.3. Occurence Rating Scale for FMEA

Score	Probability *	Modified Probability	%
10	Failure almost certain	More than Once a Day	> %30
9	Very high number of failures likely	Once ın Two or Three Days	<= %30
8	High number of failures likely	Once a week	<= %10
7	Moderately high number of failures likely	Once a month	<= %5
6	Medium number of failures likely	Once three months	<= %1
5	Occasional number of failures likely	Once in six months	<=1/10000
4	Few failures likely	Once a year	<=6/100000
3	Very few failures likely	More than once a year	<=6/million
2	Rare number of failures likely	More than once in three years	<=3/ 10 million
1	Failure unlikely. History shows no failure	More than one in five years	<=2/billion

* *Narayanagounder and Gurusami (2009)*

Detection (D): Detection is the capability of the current design control to detect the failure mode. Detection is the score indicating the ability to detect the possible failure modes or potential causes. Detection Rating Scale is indicated in Table 14.4.

Table 14.4. Detection Rating Scale for FMEA

Score	Criteria*	Modified Criteria
10	No known techniques available.	Cannot be detected or is not checked.
9	Only unproven or unreliable technique(s) available	Control is achieved with indirect or random checks only
8	Proving durability tests on products with system components installed.	Systematic sampling and inspection
7	Tests on product with prototypes and system components installed.	Control is achieved with manual Inspection for all units.
6	Tests on similar system components	Manual Error –Proofed Systems
5	Tests on preproduction system components	Control is achieved with charting methods such as SPC (Statistical Process Control)
4	Tests on early prototype system elements	SPC is only used for out of control situations
3	Simulation and/or modeling in early stage.	SPC is only used for out of control situations and all unıts have values between control limits
2	Proven computer analysis available in early design stage	Control is checked with automatic inspection system
1	Proven detection methods available in concept stage	Error detection and immediate intervention is possible, Remote to effect customer satisfaction

* *Narayanagounder and Gurusami (2009)*

Initially the process design begins with the topics that have the highest RPN score and recommended actions are performed in a decreasing order. Like all the other projects, limited sources that is planned to use may be the boundaries for all performed actions. Therefore, it is appropriate to rank RPN scores as shown in the Table 14.5 in order to classify the process requirements and to create our action plan for the company benefits.

RPN scores table indicating the risk levels for different scores is mainly separated into the distinct layers for prioritization (Table 14.5). Hence, we made the similar classification in our study with the RPN tables explained in the literature.

Table 14.5. RPN prioritization table

RPN	Reaction level	Description
0-200	Very Low Probability	Unlikely to happen and low effect on issue. It may not be effective even if the relevant subject is not studied on
201-300	Low Probability	Unlikely to happen and low effect on issue, however it may decrease the process maturity level(capability) if the necessary actions aren't taken
301-400	Moderate Probability	Probability is at % 50 level so that it may lead to serious problems for the process capability
401-500	High Probability	Level of probability that the problem to be solved, likely to happen if the measures are not taken
501-...	Very High Probability	Measures need to be taken immediately, otherwise it may affect the operations of the process

14.2.4 Alliance Management Process Design with FMEA

In this study, potential failure modes, possible causes and process requirements are determined by FMEA analysis. An example calculation is shown in Table 14.6 Moreover, process requirements are defined and prioritized by calculating RPN scores which are used to make the process design. For example, in the first process step which is the formation of the need for alliance, potential failure mode is determined as "Unable to create the suitable environment for the alliance". The potential effect of this failure is examined as "Unable to establish an effective alliance" and the severity scale of this situation is assigned as 8. The potential cause of this situation is determined as "The absence of a corporate approach for alliances" with the occurrence scale of 9. Since a new system is designed in this study, the detection step cannot be applied. Hence the detection rating scale is assigned as 9. The RPN score of this process, according to severity, occurrence and detection scales is calculated as 648. Lastly, the recommended action against this potential failure mode is determined as "To develop a corporate approach for alliances". The RPN scores of the remaining process steps of Alliance Management Process Design are calculated by this way.

Table 14.6. Alliance Management Process with FMEA

Process	Process Steps / Function	Potential Failure Mode	Potential Effect(s) of Failure	Severity (S)	Potential Cause(s) of Failure	Occurrence (O)	Current Process Cont.	Detection (D)	RPN	Recommended Action(s)
ALLIANCE MANAGEMENT PROCESS	The formation (appearance) of the need for alliance	Unable to create the suitable environment for the alliance	Unable to establish an effective alliance	8	The absence of a corporate approach for alliances	9	None	9	648	To develop a corporate approach for alliances
	Deciding the alliance	Not a well – defined issue of alliances	Establish a non – value added alliance	6	The absence of an expert board to define the issue of alliances	7		9	378	To create an expert board for alliances including proficient members
		Undefined types of alliances	Unable to manage collaborations on specific issues	6	The absence of the definite standard types of alliances	9		10	540	To classify the alliances and standardize each different types
		Choosing the wrong form of co-operation	Unable to manage collaborations on specific issues	6	The lack of a defined procedure for selecting alliance	8		10	480	To identify the procedure for selecting alliance
	...	...	...	...	...	...		...	...	...

14.3 Conclusion

In this study, we proposed an integrated administrative model designed for the alliance management process. Firstly, we determined potential failure modes on every process steps for operations. These failure modes mentions the process deficiency which we will define our solution methods to develop our business process. As a result of FMEA the recommended actions are ranked according to RPN scores which enable us to prioritize our subjects. Thus, actions scored more than 500 classified as "very high probability" have to be taken immediately and we carry on our action plans with the topics between RPN scores of 401 and 500 which are classified as "High Probability". Also "Moderate probability "actions must be taken into consideration which have the RPN scores between 301 and 400. We will perform all the recommended actions since all the RPN scores are more than 300 which we have the group of moderate probability. An example implementation order of design as a result of our FMEA study is indicated in Table 14.7.

During the implementation phase similarity of jobs and determined priority are also taken into consideration. However it is not appropriate to identify the final procedures before the configuration of all the process steps by performing the required actions. Therefore, identifying the final procedures is scheduled in the future studies although this action has one of the highest RPN scores. Moreover the complementary and similar actions are considered to have the same priority and to be performed at the same time.

Table 14.7. FMEA result table

Potential Cause(s) of Failure	RPN	Recommended Action(s)	Implementation	Imp. Sequence
The absence of corporate approach	648	To develop a corporate approach	To determine the concept of alliance type and objectives	1
Absence of a systematic method to choose the best among the alternatives	560	To establish multi-criteria decision-making technique to choose among the alternatives	To establish multi-criteria decision-making technique to choose among the alternatives	2
The absence of performance evaluation system	560	To create performance evaluation structure for alliance	To establish the performance evaluation structure of alliance with Balanced Score Card	3
Untrained staff on the structure of alliance inventory	560	To add the current and potential partners to inventory of alliance	To design the firm inventory structure	4
...	...	...	...	...

In conclusion, risk for potential failures are identified by using FMEA thus leading us to configure a strong structure. In thıs study, FMEA is used for process design. The process designed in this study is applicable for all corporations. It is implemented in a textile company and a significant progress is observed in a short time.

This study may be improved by taking the actions that is provided by FMEA, and alternatives may be simulated. As a further study, a portal application and multicriteria decision making model may be applied in the advancing stages.

Acknowledgement

"This work was supported by Research Fund of the Erciyes University. Project Number: FYL-2013-4687."

14.4 References

Alexieva, V. N. (2012) Exploring the state of business processes management in the Bulgarian enterprises. Procedia - Social and Behavioral Sciences 62: 1350 – 1354

Aras, A. A. (2005) Sustainable Process Management (in Turkish). KalDer, Istanbul

A.S.A.P. (Association of Strategic Alliance Proffessionals) (2002) Strategic Alliance Best Practice User Guide, The Warren Company

Bamford, J. D., Gomes-Casseres, B., Robinson, M. S. (2003) Mastering Alliance Strategy: A Comprehensive Guide to Design, Management, and Organization, 1st edn. John Wiley & Sons, San Francisco, pp.12

Chen, L. H., Ko, W. C. (2009) Fuzzy linear programming models for new product design using QFD with FMEA. Applied Mathematical Modelling 33: 633–647

Davenport, T. H. , Short, J. E. (1990) The New Industrial Engineering: Information Technology and Business Process Redesign. Sloan Management Review 31(4):13-14

Duysters, G. (2001) Partner or Perish: Surviving the Network Economy. Technische Universiteit Eindhoven 1-30

Jeston, J., Nelis, J. (2006) Business Process Management. Elsevier, USA, 439

Johnson, K. G., Khan, M. K. (2003) A Study into the use of the Process Failure Mode and Effects Analysis (PFMEA) in the Automotive Industry in the UK. Journal of Materials Processing Technology, 139: 348-356

Narayanagounder, S., Gurusami, K. (2009) A New Approach for Prioritization of Failure Modes in Design FMEA using ANOVA. World Academy of Science, Engineering and Technology 25

Osborn, R., Hagedoorn, J. (1997) The Institutionalization and Evolutionairy Dynamics of Interorganizational Alliances and Networks. Academy of Management Journal 261-278

Sellappan, N., Palanikumar, K. (2013) Development of Modified Evaluation and Prioritization of Risk Priority Number in FMEA. International Journal Of Engineering (IJE), 7(1)

Varadarajan, P., Cunningham, M. (1995) Strategic Alliances: A Synthesis of Conceptual Foundations. Journal of the Academy of Marketing Science 282-296

Chapter 15

An Approach to Integrate Data Mining into the Development Process

R. Lachmayer and P. Gottwald

15.1 Introduction

The development of modern technical systems focusses smart products for a better energy efficiency and handling for the user. They are increasingly equipped with intelligence. In the Collaborate Research Centre (CRC) 653 "Gentelligent Components in their Lifecycle" manufacturing technologies and materials for realization are currently being developed (Denkena *et al.*, 2010). A lot of measurement technique is implemented to realize such smart products. This opportunity leads to challenges for product development. One of them is the appropriation of life cycle data for the development process (Abramovici *et al.*, 2012).

The gentelligent technology investigates a further research effort. Principles of biology are transmitted to the applied and developed methods and tools (Lachmayer *et al.*, 2012a). In this "algorithmic design evolution based on product lifecycle information" subproject of the CRC a technical inheritance process is developed. In contrast to the autogenetic design theory (Clement, 2005) and the nature-inspired process model (Parvan *et al.*, 2012) this research observes the process of an intergenerational product life cycle. In this process the life cycle experience of the product for an adaption of the product design of the next generation is considered (Lachmayer *et al.*, 2013). Illustrating this process model the definitions of the methodology and a process model of the product life cycle which includes data mining methods have to be clarified.

Afterwards the scope of this process model has to be defined. By analyzing the requirements of this process model the tools and visualization methods have to be extracted (Eckert *et al.*, 2010).

In this case the technical inheritance for an intergenerational process model and the associated semantics are in focus. The approach how to integrate methods of data mining into the process of development is described. The example of a wheel suspension demonstrates the new opportunities for a targeted feedback and its influence for the development of the next generation.

M. Schabacker et al. (eds.), *Modelling and Management of Engineering Processes*, DOI 10.1007/978-3-662-44009-4_15

15.2 Methodology

Present research investigates the algorithmic feedback of product life cycle information for design evolution. This process is developed considering a technical inheritance. The applied semantics for this methodology are necessarily defined as follows.

Technical Inheritance:
Technical inheritance means the further development of the next generation of components or technical systems considering the experiences during the product life cycle of previous generations.

Design Evolution:
The adaption of products by analyzing product life cycle while taking evolutionary mechanism into account is defined as design evolution.

Algorithmic Data Feedback:
An algorithmic data feedback means a goal-orientated monitoring of products implementing methods of data mining in order to extract life cycle data. These data are attributed to the results of the development process of the next product generation.

Gentelligent Component:
A mechanical / mechatronical component which is featured to collect, save and transmit product life cycle information is defined as gentelligent component.

The algorithmic data feedback occurs in three different types. The direct form, which means a directly parametric correlation of life cycle information and product modelling, and the iterative form, which implements complex correlation between product model and life cycle data (Sauthoff *et al.*, 2013). Moreover the indirect form, which requires a data management for further information of the product experiences, will be analyzed in further research steps.

This research project is divided into three packages. The first is referred to as "Design of Gentelligent Systems" and includes the investigation of the methodology for a targeted feedback process. The second aspect "Statistic Operator" operates with data mining methods to transform usage data of the product into life cycle information. The third investigation "Design Optimization" deals with the subject of optimization strategies.

The first package and the contained investigations in relation to the methodology are in focus in this paper. The challenge of a targeted feedback is analyzed by the selection of the measurement technology which is integrated in the product. The options are dependent on the product application and the desired information. It is necessary to clarify which information is required for design evolution.

The principle approach of described project is depicted in Figure 15.1. In the pictogram the phases of development processes are positioned on the right whereas the usage phase of products is illustrated on the other side. Furthermore the technical inheritance couples these phases of product life cycle including the three described research packages.

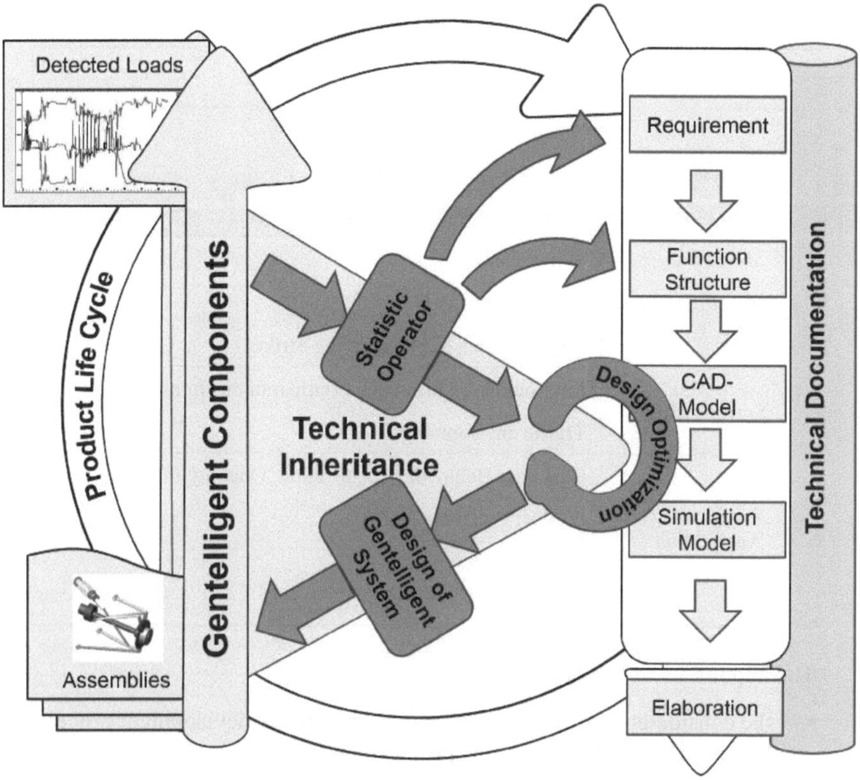

Figure 15.1. Principle approach of technical inheritance

Life cycle information is divided into four aspects by the contemplation of the development procedures of Roth (Roth, 2000) and VDI 2221 from "The Association of German Engineers" (VDI 2221, 1993). In Table 15.1 the aspects with associated examples of information are depicted.

These kinds of life cycle information are detected by different measurement techniques. For example there are conventional strain gauges or customer interviews. The measurement techniques implicate two problems for the evaluation of the data sets. At first the classification of the measuring mode influences the results. There are existing differences between discrete and analogue and also between continuous and discontinuous measured data. The second influence represents the type of data. This could be a failure information or the temperature characteristic of a component. Beyond the data sets have to be interpreted for the usage in the development process. Therefore a lot of mathematical options are announced to transform these data into information for the developer.

An additional aspect represents the gentelligent technology. These components detect the loads during their life cycle inherently.

Therefore component-specific strategies have to be compiled for a transformation of the inherent data into useful data for the next generation's development process.

Table 15.1. Product life cycle data

Aspect	Information
Physics	Kinematics Wear, Loads Energy
Life Cycle Incident	Maintenance Manufacturing Single-Events (e.g. Lightning Strike)
Surrounding	Environmental Influence (Temperature, Humidity, …) Traffic environment
Application	Customer Behavior (Localization, Quantity of Usage, …) Product Behavior Communication Ergonomics

The hypothesis of this background are:

- the consideration of methods of data mining in the development process facilitate the monitoring of the product life cycle;
- by the integration of data mining methods for a technical inheritance the view changes to intergenerational development processes.

15.3 Integration of Data Mining into a Development Process

At the beginning of the integration the scope of this model has to be analyzed. The modelling goal is classified into three aspects: The model type, the application and the system. Following from this the boundaries and parameters have to be identified (Cameron *et al.*, 2011). Due to these steps the requirements for this process model are derived and an illustration was determined.

This type represents a continuous model in consideration of technical inheritance. In this research gentelligent components are in focus. It follows that the application implies associated physical affiliations of the mechanical components and their design as well as included optimization strategies for the adaption of the next component generation. From system view this model is based on biological processes.

The scope of this process model is to represent new opportunities for different users to integrate targeted monitoring concepts into their development process. In this case the boundaries were set for mechanical components. Therefore the identified parameters are the mechanical loads during the component's product life cycle. Possible parameters are product-related forces or the temperature characteristic.

With this information it is possible to select the best-fit representation model of the process as well as the analysis of the subsequent requirements are being identified.

Development processes can be illustrated by flowcharts or precedence diagrams. In this case the challenge of the technical inheritance is to integrate methods of data mining into the intergenerational product life cycle. The requirements for this process model are:

- steps and methods of development process;
- different stations within the product life cycle;
- operations of data mining.

These activities are implemented in a model of the technical inheritance. Therefore activity diagrams of the Unified Modeling Language (UML) which are mainly used in software development are applied. In Figure 15.2 the elements of the UML are characterized which are appropriated in process descriptions. This modeling language disposed some benefits. Each activity in the process implies a consequent result, like e.g. after designing activity a CAD-Model or after assembling the real product. Additionally the activities can be grouped in different layers by using parent activities which unite some process activities. Another aspect is the opportunity to link knowledge data bases with different activities to the process. Even in times of parallelization by splitting and synchronization elements activities could perform simultaneous.

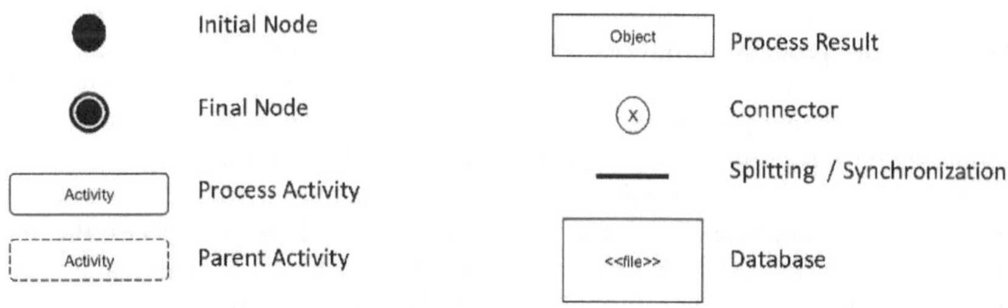

Figure 15.2. Legend of used UML

Furthermore analysis of different modelling language have shown the preference of the application of UML for process illustration (Lachmayer *et. al.*, 2012b).

Compared to traditional life cycle specifications this process is extended to a targeted product monitoring and to the feedback of life cycle information into the next generation's development process. An overview about the product life cycle considering a technical inheritance is depicted in Figure 15.3.

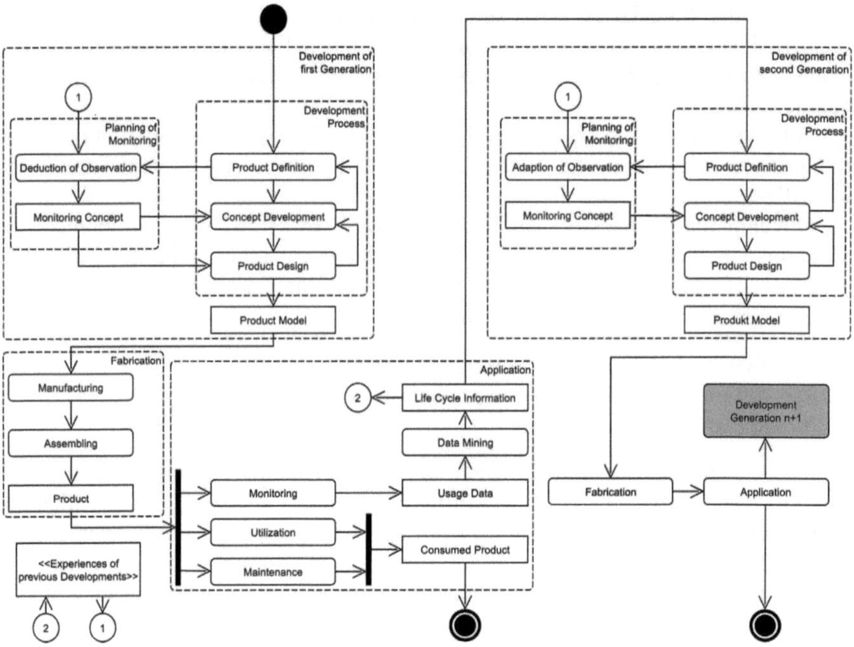

Figure 15.3. Product life cycle including technical inheritance

At first an analysis of different approaches for the development of technical systems has to be done. Focus of this investigation is the influence of life cycle data to the results of different steps of development process. The next steps are to integrate the process of planning of monitoring concept. This process has to take place after the product definition. This is necessary because the selection of the monitoring influences the steps of concept development and product design by the applied measurement technique. Another impact for the suitable observation of the product represents previous developments and associated experiences.

During the usage of the product the life cycle data are recorded continuously or discretely. This application of the measurement technique influences the data mining method. The significant information could be detected by using appropriate data mining. This life cycle information has to be saved in knowledge repositories which can be used for the next generation's development process. The application of this life cycle data in the different phases of product development depends on the type of algorithmic data feedback.

Exemplarily this approach for the technical inheritance is applied to a wheel suspension of a race car. This race car was developed by students from Leibniz Universität Hannover and symbolizes one of the demonstrators in the CRC.

15.4 Application

Monitoring operations are used in different domains. For example the structural health monitoring prevalent in traffic analysis or remote monitoring for network services. Another example for the application of monitoring and the consequentially information can be the assistance of the X-in-the-Loop methodology in product development (Düser, 2010).
In this case the life cycle monitoring is focused. Default industrial software tools offer the customer different information about machine position, usage time or loads. This information supports the customer in safety and working efficiency aspects. Beyond this information assists service teams in maintenance planning for reliability.

This approach implements product-specific monitoring concepts with an associated data mining method for mechanical components. After specifying product details the concept has to be declared. In the initially mentioned wheel suspension four components were identified for the technical inheritance. In Figure 15.4 the assembly with the specified parts is depicted.

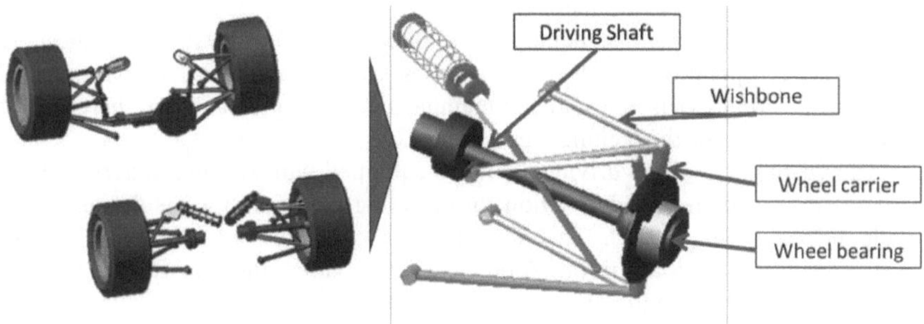

Figure 15.4. Components for monitoring concepts

At the example of the wheel carrier the selection of the monitoring concept and the data mining method is explained. First of all the design goals were derived from the product requirements. In case of the wheel suspension part of the chassis ensues following influencing aspects (Braess *et al.*, 2013):

- forces;
- safety;
- designed space;
- weight;
- ergonomics;
- ride characteristics, etc.

Based on these aspects the requirements of the monitoring concept are evaluated. In automotive industry one of the major goals is the light weight design. Moreover, for safety-relevant components an uniform load distribution is desired. These goals are connected to the optimization strategy for the adaption in the technical inheritance. For a better understanding what the components expire during their life cycle the monitoring parameters were detected. In this case for the wheel suspension the forces are from particular interest. This information is linked up with the associated components.

With load information about the wheel bearing the interval of maintenance is predictable more accurately. This flow is classified in the indirect data feedback. There exist some approaches like the research "Component status driven maintenance" (van Thiel *et al.*, 2010).

The wishbone and driving shaft are characterized in the direct data feedback. With the information about the expired forces the design is adaptable. This is realized by the linkage of the stresses and the diameters of the component.

The wheel carrier and its information about the expired loads is part of the iterative data feedback. This component consists of four application points. (Application point means the linkages between the wheel carrier and the other components of the wheel suspension.) In this case different load cases have to be considered for the design evolution.

After characterizing the types of the algorithmic data feedback and the definition of the monitoring parameter for the wheel suspension the deduction of the observation has to be defined.

The wheel bearing is equipped with laser structure strain gauges. By mathematical operations the expired stresses were transformed to loads. In this case this information is relevant for maintenance strategies and are therefore not further examined for the technical inheritance.

The wishbone and the driving shaft are equipped with sensing materials. Laser structured sensors in the edge region for detecting the highest loads during the life cycle (Mroz *et al.*, 2012). This data type is characterized in section discrete measurement. The load information which provides the maximum stress is associated directly with the diameters of the shaft.

The wheel carrier is manufactured from magnetic magnesium. With the application of an eddy current sensor the change of the magnetic field are detected at defined cross sections. This measured magnetic field modifications correlates to the forces engaging on the material (Klose *et al.*, 2012). Furthermore this component is equipped with conventional strain gauges at the position of highest expected elongation. With this information the component is adaptable to its real life cycle use by the design evolution. Information about these monitoring technologies is stored in a design catalogue. This catalogue contains information about conventional methods to observe mechanical components as well.

With the selection of the observation of the wheel suspension the monitoring concept is identified. In the next step the adaption of this concept design guidelines for implementing the monitoring technologies into the component has to be found. Also the mathematical operations for transforming the life cycle data for the developer as well as the extraction of the necessary data mining method has to be defined.

The design guidelines for each technology are also stored in the design catalogue. Information from previous developments also influences the selection of suitable data mining methods. The inherently measured data of the wheel carrier was transformed by a component-specific transformation model. Therefore the elongation is recalculated to the applied forces at the application points. It is necessary to declare a data mining method to consider different load cases.

Therefore the method of load case characterization was identified. This method has the opportunities to find quantity of loads as well as to fade out uninteresting load cases like data sets from car standing. The data sets of the wishbone and the driving shaft are directly coupled with dimensioning formulas. By linking the monitoring concept with data mining methods the integration of this process in the steps of development and targeted feedback of life cycle information is realized. The procedure of this analysis to assist the development process for the technical inheritance is depicted in Figure 15.5.

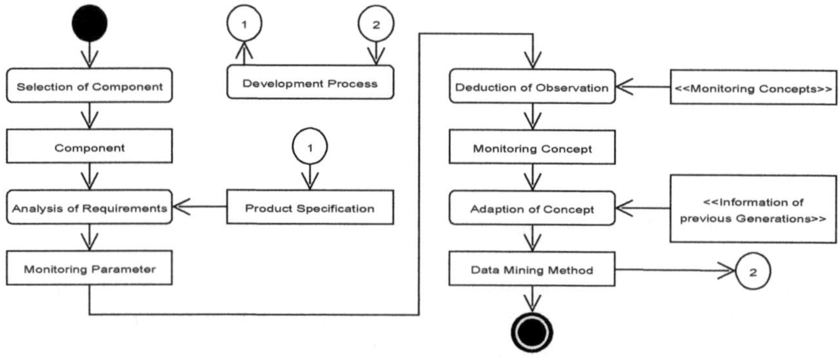

Figure 15.5. Procedure to implement useful monitoring concept

With the application of the wheel suspension the integration of data mining methods into the development process is demonstrated. The selection of the appropriate data mining method offers the advantage that only develop-relevant data are measured. This information is inherited to the next generation. By this way this approach facilitates the monitoring of the whole life cycle.

Cause this implementation of the technical inheritance the scope of product life cycle changes. To illustrate this process it is necessary to consider an intergenerational product life cycle.

15.5 Conclusion

The definition of the applied semantics forms the base for the three presented research packages in this project. In order to implement a technical inheritance of products it is necessary to select an appropriate monitoring concept. Therefore the gentelligent technology is an innovative approach to collect life cycle data. With the expanded model of the product life cycle the dependencies of the methods of data mining and the selection of the monitoring concept are shown. Also the influence of the life cycle information to the next generation's development process is illustrated by the wheel suspension. Beyond the view of an intergenerational product life cycle is depicted in Figure 15.6.

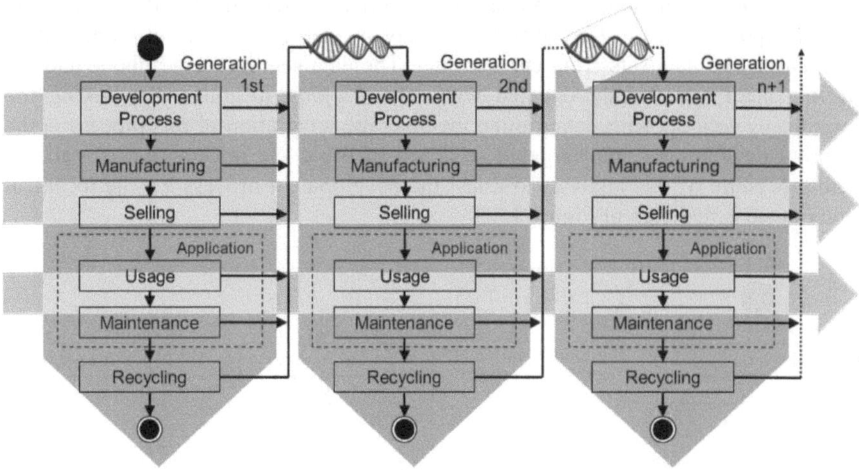

Figure 15.6. Technical inheritance in intergenerational view

This model of the technical inheritance is similar to the validation model of the VDI guideline 2221. The difference is that the technical inheritance implies the adaption of components based on real life cycle information.

The integration of data mining methods has been demonstrated with associated monitoring concept realizing a targeted feedback of life cycle data. The view of product life cycle was expanded, too.

Further research investigates the relationship between individual measurement techniques and the desired life cycle data. Also methods of data mining are analyzed to generate the significant information from the life cycle data. The scope of this research project is a development environment for the technical inheritance including methodology, knowledge bases and optimization strategies for product-specific challenges.

Acknowledgments

The results presented in this paper were obtained under the umbrella of Collaborative Research Centre 653 „Gentelligent Components in Their Lifecycle“, preliminary inspection project N4. The authors would like to thank the German Research Foundation (DFG) and the CRC 653 for its financial and organizational support.

15.6 References

Abramovici, M., Fathi, M., Holland, A., Neubach, M. (2008) PLM-basiertes Integrationskonzept für die Rückführung von Produktnutzungsinformationen in die Produktentwicklung, Werkstatt-technik online, Heft 7/8, pp. 561-567

Braess, H.-H., Seiffert, U. (Hrsg.) (2013) Vieweg Handbuch Kraftfahrzeugtechnik, ATZ/MTZ-Faxchbuch, 7. AuflageSpringer Vieweg Verlag

Cameron, I., Gani, R. (2011) Product and Process Modelling, A case study approach, first Edition, Elsevier

Clement, S. (2005) Erweiterung und Verifikation der Autogenetischen Konstruktionstheorie mit Hilfe einer evolutionsbasierten und systematisch-oppertunistischen Vorgehensweise, Dissertation, Otto-von-Guericke-Universität, Magdeburg

Denkena, B., Henning, H., Lorenzen, L.-E. (2010) Genetics and intelligence: new approaches in production engineering, Production Engineering, Volume 4, Number 1, pp. 65-73

Düser, T. (2010) X-in-the-Loop – ein durchgängiges Validerungsframework für die Fahrzeugentwicklung am Beispiel von Antreibstrangfunktionen und Fahrerassitenzsystemen, Forschungsberichte IPEK, Dissertation, Band 47

Eckert, C. M., Stacey, M. K. (2010) What is a Process Model? Reflections on th Epistemology of Design Process Models, Proceedings of the 1st International Conference on Modelling and Management of Engineering Processes 2010, Springer Verlag, pp. 3-14

Klose, C., Mroz, G., Kerber, K., Reimche, W., Bach, F.-W. (2012) Production and Comparison of Adapted Load-sensitive Magnesium Alloys, Procceedings 1st Joint International Symposium on System-integrated Intelligence: New Challenges for Product and Production Engineering, pp. 56-58

Lachmayer, R., Sauthoff, B., Gottwald, P. (2012a) Product Optimization by Analysis of Gentelligent Life Cycle Information, In Proceedings of SysInt, 1st Joint International Symposium on System-Integrated Intelligence 2012: New Challenges for Product and Production Engineering, pp. 28-31

Lachmayer, R., Gottwald, P. (2012b) Evolutionäre Produktentwicklung – Integration von Nutzungsinformationen, Innovationspotentiale und Perspektiven von CAE-Entwicklungsumgebungen, Kolloquium Produktentwicklung: Band 3, PZH-Verlag Garbsen, pp. 1-7

Lachmayer, R., Mozgova, I., Deiters, A., Sauthoff, B., Gottwald, P. (2013): Design Evolution by Integration of Life Cycle Information in Knowledge Repositories, Consilience and Innovation in Design, 5th International Congress of International Association of Societies of Design Research, Tokyo, Japan, Vol. 2, pp. 309-318

Mroz, G., Reimche, W., Bach, Fr.-W. (2012) The use of component's edge region as inherent information carriers and loading indicators, Procceedings 1st Joint International Symposium on System-integrated Intelligence: New Challenges for Product and Production Engineering, pp. 158-160

Parvan, M., Miedl, F., Lindemann, U. (2012) Nature-inspired Process Model for Concept Selection and Evaluation in Engineering Design, NordDesign, Aalborg, Denmark, pp. 486-493

Roth, K. (2000) Konstruieren mit Konstruktionskatalogen, Band I, Springer Verlag

Sauthoff, B., Lachmayer, R. (2013) Optimierung von Bauteilen auf Basis von Produktnutzungsinformationen, Beitrag zum 24.DfX Symposium, Jesteburg, pp. 119-130

Van Thiel, B., Nyhuis, P (2010) Datenbank zur gentelligenten Instandhaltungsplanung, ZWF Zeitschrift für wirtschaftlichen Fabrikbetrieb, 4/2010, pp. 339-343

VDI 2221 (1993) Methodik zum Enmtwicklen technischer Systeme und Produkte, Beuth-Verlag

Chapter 16

Optimal Sceduling of Stochastic Production Processes Through Model Checking

L. Herbert, Z. N. L. Hansen, R. Sharp and
P. Jacobsen

16.1 Introduction

Modelling and subsequently optimising workflow processes has been a key part of efforts to improve efficiency in production and engineering firms since the beginning of the 20th century (Gilbreth & Gilbreth, 1921). These developments are called *Business Process Modelling* (BPM), which is a discipline concerned with the mapping of business workflows, for example in production, to enable analysis and improvement of organisational efficiency and quality. However, modern enterprises, in particular those involved in producing highly engineered products or addressing dynamic customer needs, are often characterized by business processes which exhibit complex concurrent behaviour incorporating unreliable or unpredictable components, and which can be subject to various non-functional requirements (for example reliability, safety or legal requirements). These properties make them difficult to model and even more difficult to optimize.

Ensuring that such complex systems are both dependable and efficient poses a significant challenge, combining the need for verification of safety properties while simultaneously requiring a specific performance profile. To achieve these goals, such systems often require sophisticated scheduling especially when the system involves stochastic elements (i.e. elements whose specific behaviour cannot be predicted in advance). Being able to synthesize a schedule for the optimal execution of such systems early in their design phase allows for accurate determination of how the system will be employed in practice. Consequently this holds the potential for the early identification and exclusion of workflow configurations for which an efficient schedule cannot be developed. This can for example help organizations introduce new technology to existing workflows, optimize existing workflows or create optimal new workflows with less effort, time and cost spent.

M. Schabacker et al. (eds.), *Modelling and Management of Engineering Processes*, DOI 10.1007/978-3-662-44009-4_16

16.1.1 Contribution

In previous work (Herbert & Sharp, 2013b), methods that allow for analysis of models of business processes expressed in the *Business Process Model and Notation* (BPMN) modelling language (Object Management Group, 2011) were developed. The properties to be analysed were described using an extended form of the temporal logic *Probabilistic Computation Tree Logic* (PCTL) (Aziz *et al.*, 1995). This approach allows for the precise calculation of the occurrence and ordering of events and their associated timing, transient and steady state probabilities of events, and costs properties.

The approach taken to schedule generation under complex constraints is to employ these earlier developed model checking methods, to explore possible resolutions of non-determinism within a BPMN model. In the approach shown in Figure 16.1 we allow for a BPMN model to be annotated with quantitative data and extended with stochastic behaviour so as to capture real world business process behaviour. Then, given a quantitative goal and a set of possible decisions, we derive the schedule, a sequence of decisions, that best approaches the goal.

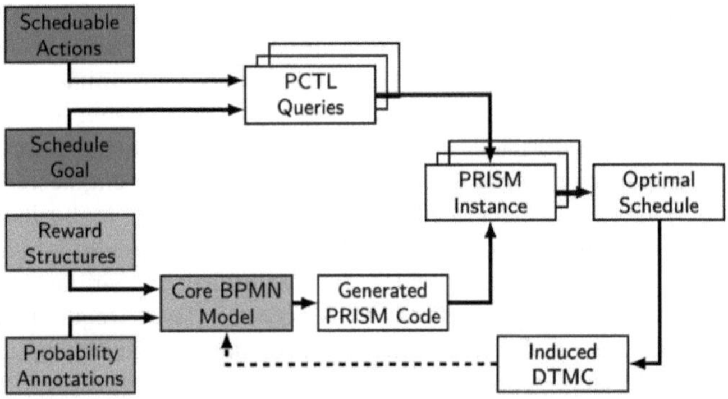

Figure 16.1. Overview of schedule generation via PRISM model checking (grey boxes mark additional inputs needed).

Schedule generation is made possible by performing model checking on specific permutations of the set of actions possible in a model to generate *optimal adversaries* which optimise (minimise/maximise) a reward value, while observing constraints which encode any required safety properties. By evaluating the quantitative properties of the generated adversaries we are able to construct an execution schedule that best approaches the desired performance properties.

While the focus of the paper is the generation of execution schedules from stochastic BPMN models, BPMN is limited by an imprecise and incomplete semantic definition. We therefore begin by briefly presenting a formalised variant of BPMN previously described in (Herbert & Sharp, 2012a). By translating BPMN models into *Markov decision processes* (White, 1993) described using the PRISM modelling language (Parker, 2012) we are able to employ the model checker PRISM (Kwiatkowska *et al.*, 2011) to efficiently generate and evaluate adversaries representing possible execution schedules. The choice of PRISM is motivated by the great expressivity of its PCTL query language, which makes it possible to express complex probabilistic constraints and schedule goals.

Fundamentally, our approach allows for business analysts to test a wide range of possible designs, and to readily *debug* them, before committing to a specific practice, which achieves a more effective development of the business process in question. These developments are illustrated using an example drawn from a large Danish baked goods company. In the last century, food production in the developed world has become increasingly more globalized, automated and industrialized, with fewer and larger players in the market. Baked goods, a key part of the food industry, have experienced the same development, leading to a market characterised by intense cost pressures, fierce competition, and few opportunities for differentiation. Baked goods are an example of a business which uses small batches to fulfil dynamic customer needs, has clear constraints (e.g. hygiene and freshness) and therefore need to combine flexibility, lean practices and quality to produce high quality products effectively and efficiently. The baked goods industry is therefore an ideal candidate to illustrate the practical applicability of this method.

16.1.2 Related Work

We are not aware of any previous work directly addressing the generation of schedules for systems modelled in BPMN, with or without stochastic extensions. Many approaches exist which focus on the synthesis of schedules for non-stochastic systems, often using various simplex based methods. However the stochastic case does not allow for the application of these methods.

The inclusion of rewards in BPMN models allows for the determination of ideal strategies, and whereas multiple schedules may exist, quantitative methods allow for the selection of schedules which optimise rewards of interest. In this regard, two similar approaches focused on solving scheduling problems using model checking are given by Wijs *et al.* (2009) and Basu *et al.* (2011). However, in both cases, the construction of the model from which to generate a scheduler is a manual process that requires considerable tuning.

It should be noted that workflow mining approaches (van der Aalst et al., 2003) complement the ideas presented in this paper. They are focused on extracting a workflow model from execution traces of an implemented system and provide a means to determining stochastic and timing data for an existing process.

16.2 Business Process Model and Notation

The *Business Process Model and Notation* (BPMN) language (OOG, 2011) is a widely adopted graphical notation for specifying workflows. The semantics and pragmatics of BPMN are, however, only informally defined in the relevant standards (OOG, 2011), thus leaving a number of issues open to interpretation.

There are essentially only two fundamental types of object, *nodes* and *flows*, and in this work only a small subset of BPMN, often known as the *core* subset, will be used. This consists of the eight elements found in a large survey to be the most commonly used in industry (Muehlen & Recker, 2008).The graphical elements of core BPMN are shown in Figure 16.2 and described in Definition 1 below. It should be noted that by combining several Core BPMN elements any element of the complete BPMN language can be simulated, even inclusive gateways (Christiansen *et al.*, 2011).

BPMN modelling involves composing a number of elements into a *business process diagram* (BPD).

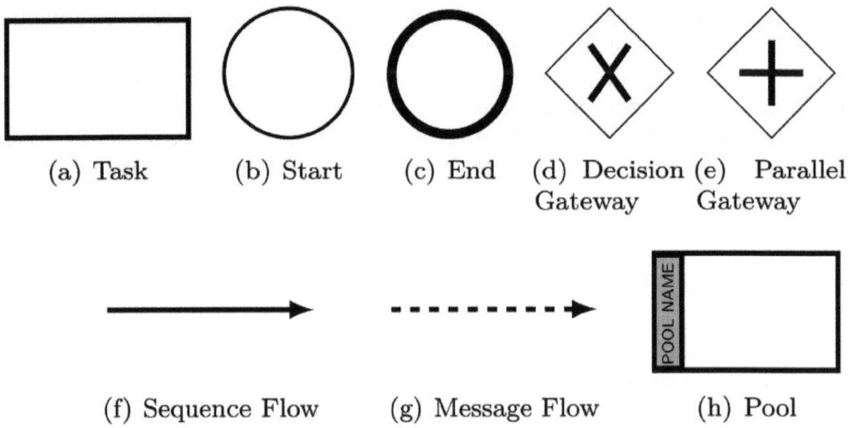

Figure 16.2. Core BPMN elements.

Definition 1 (Stochastic Core BPD). A Stochastic Core BPD is a tuple $BPD = (\boldsymbol{N}, \boldsymbol{F}, \boldsymbol{P}, pool, \boldsymbol{L}, lab, P)$ where $\boldsymbol{N} \subseteq \boldsymbol{T} \cup \boldsymbol{E} \cup \boldsymbol{G}$ is a set of nodes composed of the following disjoint sets:

- *Tasks* **T** *are the basic actions done as part of a given workflow, e.g. "sending a letter" or "putting sprinkles on a cake".*
- *Events* $\mathbf{E} \subseteq \mathbf{E^S} \cup \mathbf{E^E}$ *where the disjoint sets* $\mathbf{E^S}$ *and* $\mathbf{E^E}$ *respectively represent start and end events.*
- *Gateways* $\boldsymbol{G} \subseteq \boldsymbol{G^D} \cup \boldsymbol{G^F} \cup \boldsymbol{G^M}$, *where the disjoint sets* $\boldsymbol{G^D}$, $\boldsymbol{G^F}$ *and* $\boldsymbol{G^M}$ *respectively represent exclusive decision gateways, parallel fork gateways and parallel merge gateways.*

$\boldsymbol{F} \subseteq S \cup M$ *is a set of flow relations, where sequence flows* $S \subseteq \boldsymbol{N} \times \boldsymbol{N}$ *relate nodes to each other and* $M \subseteq \boldsymbol{T} \times \boldsymbol{G^M}$ *is a relation between tasks and parallel merge gateways.* $\boldsymbol{P} \cup \wp(N)$ *is a set of disjoint pools and* $pool: \boldsymbol{N} \rightarrow \boldsymbol{P}$ *assigns nodes to a pool,* $p \in \boldsymbol{P}$ *is a set of unique labels and* $lab: \boldsymbol{F} \rightarrow \boldsymbol{L}$ *is a labelling function which assigns labels to flows. The function* $P_g: S \times \boldsymbol{L} \rightarrow [0,1]$ *is a partial function which for a node* $g \in \boldsymbol{G^D}$ *and label* $l \in \boldsymbol{L}$ *assigns probabilities to all outgoing sequence flows* $((g,x,l))$, *such that for a given* l: $\sum_{\forall x \in out(g)} P_s((g,x)l) = 1$.

The definition of a *BPD* given in Definition 1 models workflows by using elements of ***F*** to define a directed graph with nodes which are elements of *N*. However, Definition 1 allows for graphs which are unconnected, do not have start or end elements, and are free-form or have various other properties which place them outside what is implied to be permitted in standard BPMN models. To ensure that a *BPD* describes a meaningful workflow we have developed a set of well-formedness rules (Herbert & Sharp, 2012b) which enforce restrictions on connecting elements, pool boundaries, and message passing.

The function *P* in Definition 1 allows for the modelling of probabilistic decision points in the modelling of business processes; with the intention of capturing real-world behaviour where the outcomes of complex decision within a process can appear random and are not possible to predict in advance. BPMN makes use of external conditions on decision gateways to select the outgoing flow from a decision point. These decisions are modelled by the set ***L*** and assigned to specific flows by the function *lab*. In practice, decision points in a workflow will have outcomes which depend on some inherent property of the task or on outside factors. The idea is that at a decision point an active choice is made, and then that choice results in a number of different possible outcomes. Figure 16.3 illustrates the application of *P* to a decision gateway *g*.

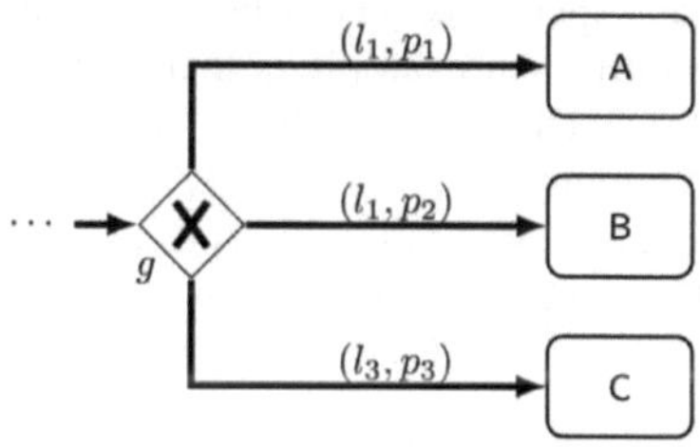

Figure 16.3. Assignment of label probability pairs to a decision gateway. Here application of *P* requires $p_1 + p_2 = 1$ and $p_3 = 1$.

To enable quantitative analysis of a workflow we add numerical data to our models by using the following function which associates positive real numbers with tasks in a *BPD*.

Definition 2 (BPD Task Reward Function). For a *BPD* a reward function for a task $t \in \boldsymbol{T}$ is a partial function $R: \boldsymbol{T} \rightarrow \mathbb{R}_{\geq 0}$.

This function captures the notion that certain nodes have some reward or cost associated with the task. We may associate as many reward structures as we wish with a given *BPD*, so that a single task may have multiple different numerical properties which are incremented when the task is performed. Further details of these structures and model checking of these properties can be found in (Herbert & Sharp, 2012a).

16.3 Case Study Details

The application of these methods was explored in a case study involving one of the largest Danish producers of baked goods which, for reasons of anonymity, is designated Baked Goods A/S. Baked Goods A/S is a business to business company which was established in 2000 and entered the market in 2001. In 2012 Baked Goods A/S had 103 full time employees. The company focused on increasing its revenue on the domestic market and on developing new export markets. Revenue increased from DKK 159 million in 2011 to DKK 180 million in 2012. The increase primarily derived from the domestic market despite the general downward price development on the Danish market. Revenue growth primarily derived from increasing sales of convenience products, but also from coffee bread products. Baked Goods A/S seeks to have a continued price focus and a high innovation level.

The company is mainly focused on differentiation by making their bake-off products appear more "home-made" by making them less regular (e.g. not completely the same size or shape), thereby introducting a controlled amount of stochasticity into the production line. However, they also produce products with strict requirements for regularity for customers for whom this is vital. As a part of their differentiation strategy Baked Goods focuses on signaling the connection to home-made, fresh and natural products; this is done through labels on the products, promotional material and how the company presents and views itself (i.e. a company located in the country-side with a focus on traditional baking values and techniques). Baked Goods experience very volatile commodity prices which makes increased efficiency a priority. An example is that a sausage roll can only be sold by the firm for around 1-2 DKK while shops take 6-8 DKK for this bake-off product, i.e. a markup of around 75%.

Baked Goods A/S have two production lines. Line 1 develops cakes and pastries and line 2 develops baked goods like sausage rolls and pizzas.

16.3.1 BPMN Model of the Case Study

Using the formalised version of BPMN defined in Definition 1 we can model the two concurrent production lines using an example of three products being produced concurrently on these lines which share key pieces of equipment (see Figure 16.4).

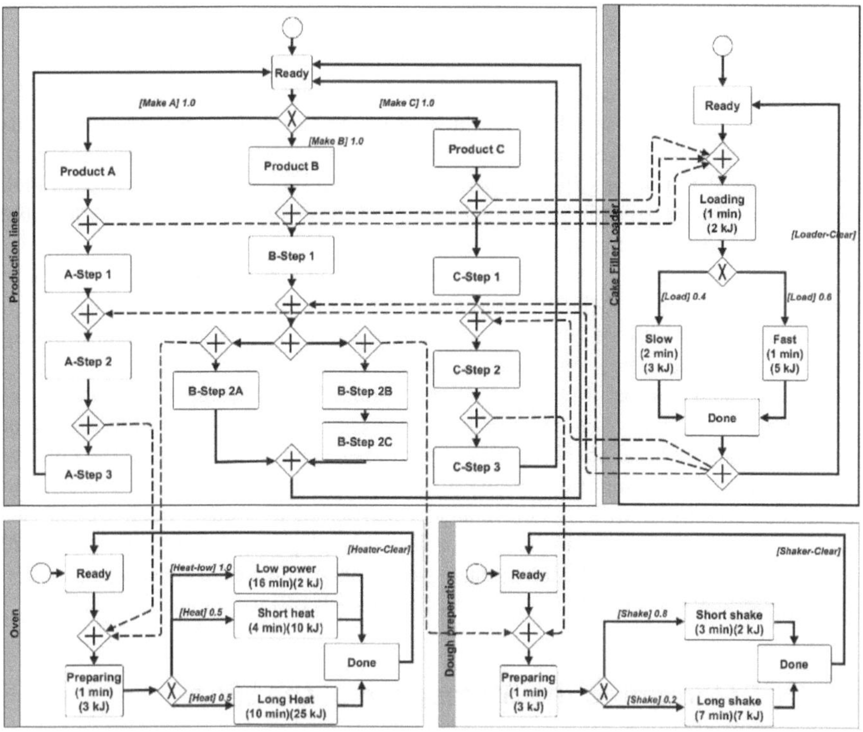

Figure 16.4. Abstracted from BPMN model of Baked Goods A/S production process

Figure 16.4 illustrates this scenario; this system consists of 4 processes each represented as an individual pool. The production lines process drives the operation of this system and makes a non-deterministic choice between producing products A, B or C. Manufacturing each product involves a specific sequence of operations performed by separate sub components; each of these performs steps that have delays which are stochastically chosen. Synchronization between the different processes (pools) is performed via the [label] constructs, in the standard fashion of BPMN. A number of states are annotated with reward structures tracking time used and energy expended.

Note that this system has two key points where a non-deterministic choice must be made between several options. Namely in the choice of which product to manufacture and, when heating products, a choice between normal or low-power heating. In this system there is a safety requirement that shaking must never occur while loading of a product is taking place, as the vibrations caused by shaking could lead the system to malfunction. The scheduling goal is that production of a specific batch of products (e.g. 2 batches of A, 1 of B and 3 of C) should be sequenced so that production takes place as quickly as possible and using the minimum amount of energy, while observing the safety requirements.

16.4 Stochastic Model Checking

The goal of this work is to make it possible to automatically transform a BPMN model of a production workflow into a form which can be formally analysed. This allows all possible execution paths through the workflow to be examined. Determining the optimal scheduling of tasks, given a set of optimisation goals, then simply involves determining the properties of each possible schedule and choosing the one which best meets the defined goals.

This is achieved by transforming the BPMN model into a *Markov decision process* (White, 1993) (MDP) which is amenable to formal statespace analysis (i.e. mathematical analysis of the set of possible values the process can take during execution). These states represent possible configurations of the system being modelled with probabilistic state transitions being combined with non-deterministic choices between several discrete probability distributions over successor states. Model checking allows for the efficient exploration of the entirety of this space with a temporal logic employed to select sets of states of interest, and offers the possibility of verifying many properties of a system. In this paper we will specifically use this capability to select sets of paths through the statespace that represent different schedules; each path is then checked to ensure that given safety criteria are observed and the values of rewards of interest are computed.

16.4.1 Determining the Statespace of the BPD

In our approach, BPMN *BPD* models are mapped directly into the guarded command language used by the PRISM model checking tool (Kwiatkowska et. Al. 2011) which allows for the efficient generation and analysis of a concurrent systems statespace. The mapping, which focuses on the control flow structure of the model, involves decomposing a BPMN *BPD* into sub-processes which are individually mapped to PRISM code, with appropriate synchronization constructs generated to maintain the same effective control flow. The central idea of the translation algorithm is to identify sub-processes of the source BPMN *BPD* and then map these to modules of PRISM code so that the encoding will be compositional and will not impose further semantic interpretation on a model than originally defined. It should be noted that PRISM will generate all possible interleaving's of two sub-processes and calculate PCTL property values for the entire statespace. An illustration of this interleaving behaviour is shown in Figure 16.5. Note how the paths from the fork to the join cover all possible allowed interleaving's.

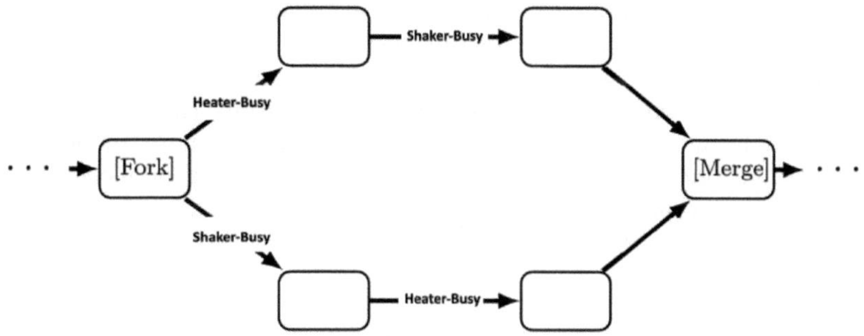

Figure 16.5. Illustration of interleaving semantics imposed by the PRISM model checker (labelled extract from Figure 16.4 of the statespace of the example from section 16.3.1)

For the complete model applying the translation processes means a statespace as shown in Figure 16.6 can be generated.

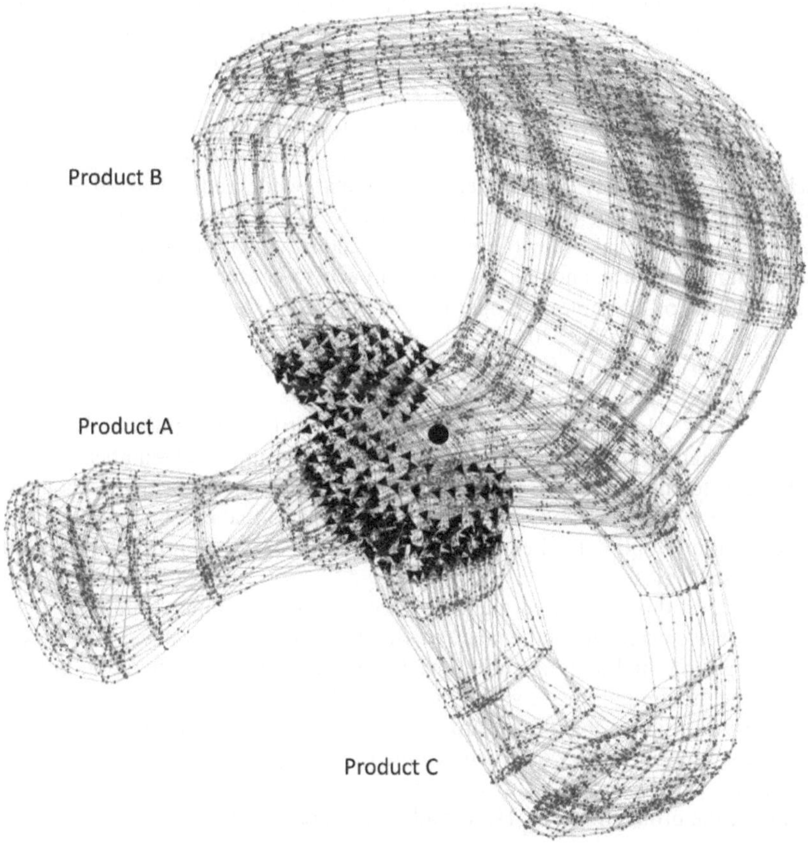

Figure 16.6. Statespace of the BPMN model of Figure 16.4 (Annotations removed, 3080 States, 10999 Transitions).

In Figure 16.6 each dot represents a unique configuration of the system which highlights the nature of this scheduling problem. The initial state is represented by the black dot and the statespace is characterised by three large loops which correspond to the manufacture of each of the three products. The high complexity of the manufacture of product B is clear in the larger number of nodes and transitions that form this loop. A solution to the scheduling problems for this system requires choosing the correct sequence of choices, marked by black triangles, to reach the scheduling goals.

16.5 Generating Optimal Schedules

Generation of a schedule requires the resolution of the points of non-determinism under the control of agents in the model. In other words in order to find an optimal schedule for a production workflow we need to be able to investigate all actions actors in the workflow may perform at any given point in the execution of the workflow. To this end we will employ PRISM's capability for adversary generation (Forejt, 2011) (i.e. generating all possible resolutions of nondeterminism) which generates an induced *discrete time Markov chain* (DTMC) (White, 1993) on the generated statespace that equates to evaluating the best or worst-case choice of actions at all decision points that satisfy a chosen PCTL constraint. We will employ Algorithm 1 to systematically explore generated adversaries that meet the scheduling and safety requirements. Note that while individual adversaries are generated on the same basic statespace, PRISM allows symmetry (Kwiatkowska *et al.*, 2006) and partial order (Gröÿer *et al.*, 2006) reduction to be employed when searching this space and allows our approach to presently scale to the feasible verification of complex properties of large systems (up to 10^{10} states (Kwiatkowska & Parker, 2012)).

In Algorithm 1, the function **ExpectedValue** computes, via model checking, the expected value of the DTMC's reward. Line 3 simply involves removing permutations where a state representing a reward having values **n+1** is required before a state where the reward has value **n**. The maximal number of possible schedules to check, and the consequent complexity of the algorithm, is bounded by product of the number and non-deterministic choices that can be made. In the case when multiple optimal schedules exist, algorithm 3 will simply return one of the optimal strategies.

Note that this simplified algorithm searches for minimum solutions, however with simple changes it can be made to search for maximum values. This is done by changing the initial assignment in line **1** to **0**, the adversary generation in line 9 to $\mathbf{R_{max}}$ (the maximum reward value), and the test in line 10 to '>'.

In line 9 multiple separate reward structures may be evaluated and a composite score for these rewards constructed. Finally, in addition to reward structures, this algorithm can be modified to employ the PCTL probability operator **P** to also optimise the probability of CTL formulas of interest being true.

Algorithm 1: Optimal Safe Schedule Selection

Input: PRISM Model, Safety Constraint set **C**, Needed Actions Set **A**
Output: Optimal Adversary X'

$X' \leftarrow \top$;
Generate **Z** the set of permutations of elements of **A**
Remove invalid permutations from **Z**
for $Z' \in \mathbf{Z}$ **do**
  String $S =$ null
  for $a \in Z'$ **do**
    $S \leftarrow S \cup (\mathsf{F}a)$
  Generate adversary X for PRISM Model under the safety constraint
  $\mathsf{R}_{min}((\mathsf{G}\ C) \wedge S)$
  if (ExpectedValue(X) $< X'$) **then**
    $X' \leftarrow X$
Return X'

16.5.1 Schedule Specification

Finally, to describe requirements for an optimised schedule, for example task A must happen before task B, we employ a mathematical language which can describe the ordering, in time, of tasks. This language is the property specification language Probabilistic Computation Tree Logic (PCTL) (Kwiatkowska *et al.*, 2011) which is based on classical continuous stochastic logic (Hansson & Jonsson, 1994) extended to probabilistic quantification of described properties. An implementation of the PCTL logic is employed by the PRISM model checker (Kwiatkowska *et al.*, 2011). While this logic allows reasoning about a wide range of system properties (Baier & Katoen, 2008), we will employ PCTL queries to filter out paths in the statespace generated by the PRISM model checker for a BPMN *BPD.*

Specifically, PCTL queries are used to define the safety properties which we require for the system and the specific tasks that we want scheduled by the system, and to determine the cumulative mean rewards values along that path. To achieve this we will restrict ourselves to the **R** operator used to express properties that relate to rewards; more precisely, the expected value of a random variable associated with a particular reward structure. Since a BPMN *BPD* will often be decorated with multiple reward structures, we augment the **R** operator with a label. For example, to query the mean time to **failure** we would specify the following property: $\mathbf{R_{time}} = \mathbf{?}$ **[F failure]**

To support analysis of the example from Section 3 we employ the following temporal CTL operators, which can be combined using traditional conjunction, disjunction and negation:

- **a U b** The binary until operator specifies that, for a given path, in some state of the path the property **b** is true and in all preceding states the property **a** is true.

- **F*a*** The eventually operator specifies that, for a given path, **a** eventually becomes true at some point along the path.
- **G*a*** The always operator specifies that, for a given path, **a** is true in all states along the path.

16.5.2 Scheduling the Case Study

To generate an optimal schedule for the case study example where the optimisation goal is to minimise time taken and energy used, we define a set of PCTL properties, and apply Algorithm 1 to determine the optimal sequence in which to produce the required products. The optimal scheduling of production has the safety constraint that loading and shaking of a product cannot take place at the same time. This safety constraint can be expressed in PCTL as: **G!([Shake] ^ [Load])**. This should be read as follows: **(G** (globally)(for all future execution paths) it is not **(!)** the case that the actions (**[Shake]** and **(^)** **[Load]** take place).

Algorithm 1 also requires a set of needed actions, **A**. Considering the example of producing 2 batches of **A**, 1 of **B**, and 3 of **C**, while also deciding when producing **A** whether to choose **low-power** or **normal** heating of the product. In our case, this set of constraints would have the following form:
$\{((A_1UH_n)U(A_2UH_n)UB_1UC_1UC_2); (B_1U(A_1UH_l)U(A_2UH_n)UC_1UC_2), \ldots\}$

Finally, we wish to generate a schedule of actions under the additional constraint that the accumulated value of the **time** and the **energy** rewards along the chosen path are the smallest possible, with equal weight being given to minimizing both rewards. This involves employing the PCTL reward operator **R** to calculate the expected value of the time and energy reward structures for the paths remaining in the set of possible paths, once the unsafe paths have been excluded.

In this case there exists a unique schedule shown in Figure 16.7 with an expected mean time to completion which is **37:4** minutes, using **98:3** kJ. In this solution the machines on the production lines chooses to begin production by manufacturing one batch of product **A** and making use of the lower power heating setting in its production. Once loading is complete for product A, manufacture of product **C** is started and repeated until the loading of the 3rd batch of product **C**. Then a second batch of **A** is started and **2:4** minutes (the mean time needed to load B) before this is completed, production of product **B** is started.

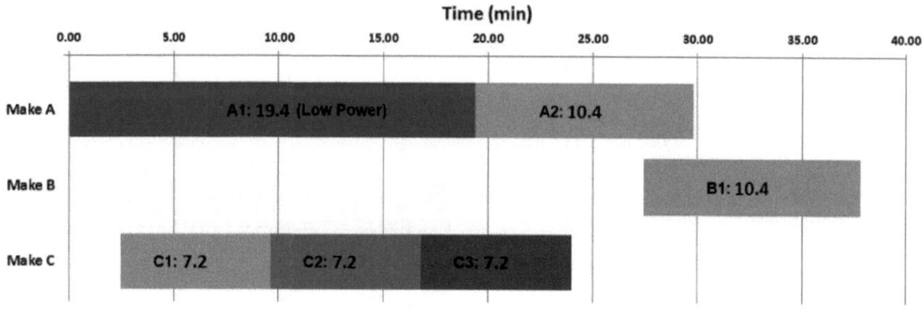

Figure 16.7. Generated minimal time/energy usage schedule for the example from Section 16.3.1 (37.4 minutes, 98.3 kJ)

16.5.3 Discussion

Existing languages for the modelling of business processes such as BPMN, UML activity diagrams or YAWL lack a formalised semantic basis which would enable formal analysis and subsequent automated scheduling. Further, these languages do not allow for modelling stochastic behaviour or provide mechanisms to effectively track the consumption of resources during execution. These aspects are therefore the key strengths of this method as no other method, to our knowledge, has all these features. Further, it should be noted that our method by employing the PRISM tool calculates exact values. However, this need for precision also means that a disadvantage of our approach is that it requires detailed knowledge of the workflows being optimised. Another disadvantage of our method is that to use the optimisation schedule in practice great computing power is needed which can be both expensive and time-consuming. However, our method allows for automatic optimal scheduling with mathematical precision and within specific parameters which can help organisations limit waste of for example energy or material as well as optimise production with regard to parameters such as time, personal and cost.

In order to use our method to optimize scheduling of production workflows in practice we have designed a prototype software tool. This tool allows practitioners to model, analyse and optimise a wide range of workflows. As the tool has a graphical GUI interface, the user does not need to have any knowledge of the technical workings of the tool. They only need to be able to associate rewards and probabilities to a workflow in order to optimise it according to the desired parameters (see Figure 16.8).

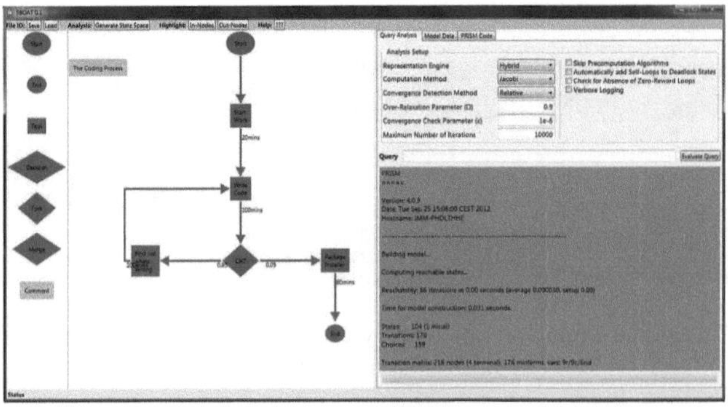

Figure 16.8. User interface for the prototype version of the software tool.

The software tool is able to model, and annotate with rewards and stochastic branching, a business process (for example in production) as a BPMN *BPD*. Analysis is specified using a PRISM style PCTL query and depending on the nature of the query one or a number of results are calculated. At the core of the tool is the PRISM model checker which performs analysis of individual models generated (see Figure 16.9).

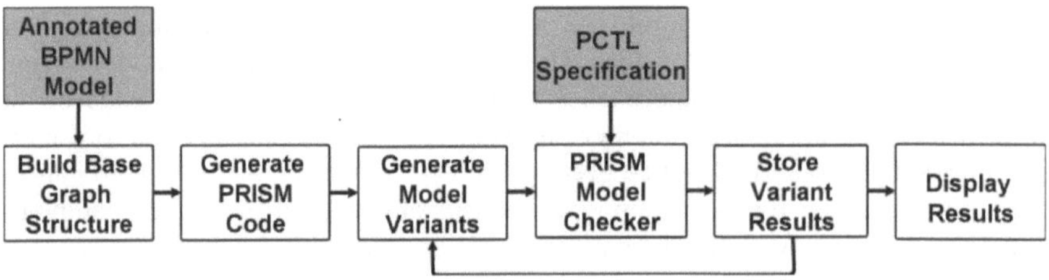

Figure 16.9: Overall design of the software tool.

For this approach and the software tool to provide the most benefits to an organisation the company need to be able to annotate their production workflows with rewards and probabilities which need to reflect the real-life scenario. This means that organisations which deal with great insecurities in their workflows or who are unable or unwilling (for example due to financial constraints) to correctly measure and annotate their production workflows will not be able to gain the full benefits of this approach. The approach presented here is most useful in large industrial settings for example in hardware manufacturing or in workflows which require precise and safe operations like in the healthcare industry.

16.6 Conclusion

In this chapter we have presented a method for model checking annotated BPMN models. This allows for the powerful quantitative analysis of such models, which can be extended to allow for the synthesis of execution schedules under multiple types of constraint. We further demonstrated the application of these methods to a simplified model of a real-world system where the solution of a complex scheduling problem can be solved with minimal effort.

Our work presents a method to allow the automatic derivation of the optimal actions a system should perform to achieve desired goals. This can be crucial in forming system design as it suggests the fashion in which a system will be employed and can help focus testing and verification efforts. When applied to existing systems, these methods can be employed to optimize the systems' behaviour.

This paper has shown the overall effectiveness and applicability of this method. The method can be used for optimisation of many other aspects, for example machine, space or personal usage. Future research should focus on improving the breath of possible candidate processes explored and computational performance.

16.7 References

Van der Aalst, WMP, van Dongen, B. F , Herbst, J., Maruster, L., Schimm, G. & Weijters, A. (2003), Workflow mining: a survey of issues and approaches, Data & knowledge engineering 47 (2), 237-267

Aziz, A. Singhal, V. Balarin, F. Brayton, R. Sangiovanni-Vincentelli, A. (1995) It usually works: The temporal logic of stochastic systems. In P. Wolper, (ed.), 7th International Conference on Computer Aided Verication (CAV'95), 939: 155-165. Springer

Ball, T., Jones, R. (2006). Eds., ser. Lecture Notes in Computer Science, London, UK: Springer-Verlag, 234-248

Basu, A., Bensalem, S., Peled, D. & Sifakis, J. (2011). Priority scheduling of distributed systems based on model checking. Formal Methods in System Design 39(3): 229-245

Baier, C., Katoen, J.-P. (2008). Principles of Model Checking. Cambridge MA, USA: The MIT Press

Christensen, M.M., Haar, S. (2013) Supermarkeder presser brødkæmpe i minus. Børsen newspaper 8/08/2013. 151:12-13

Christiansen, D. R., Carbone, M. & Hildebrandt, T. (2011), Formal semantics and implementation of BPMN 2.0 inclusive gateways, in Proc. of the 7th international conference on Web services and formal methods, ser. Web Services and Formal Methods 2010, Berlin, Heidelberg: Springer-Verlag, 146-160

Gilbert, C, Bower, J. L. (2002) Disruptive change. When trying harder is part of the problem. Harv Bus Rev. 80(5):94-101

Gilbreth, F. B , Gilbreth, L. M. (1921) Process Charts: First steps in finding the one best way to do work. American Society of Mechanical Engineers (ASME)

Gröÿer, M., Norman, G., Baier, C. Ciesinski, F., Kwiatkowska, M. & Parker, D. (2006), On reduction criteria for probabilistic reward models, in Proceedings of the 26th international conference on Foundations of Software Technology and Theoretical Computer Science, ser. FSTTCS'06, Berlin, Heidelberg: Springer-Verlag, pp. 309-320

Hansson, H., Jonsson, B. (1994), A logic for reasoning about time and reliability, Formal Aspects of Computing, 6(5): 512-535

Herbert, L., Sharp, R. (2013a) Optimisation of BPMN business models via model checking. Proc. for ASME 2013 International Design Engineering Technical Conferences and computers and information in Engineering Conference (IDETC/CIE2013)

Herbert, L., Sharp, R. (2013b) Precise analysis of probalistic business process model and notation workflows. Journal of Computing and Information Science in Engineering, 13: 2-11

Herbert, L., Sharp, R. (2012a) Quantitative analysis of probabilistic (BPMN) workflows. Proc. for ASME 2012 International Design Engineering Technical Conferences and computers and information in Engineering Conference (IDETC/CIE2012)

Herbert, L. & Sharp, R. (2012b), Using stochastic model checking to provision complex business services, in High-Assurance Systems Engineering (HASE), IEEE 14th International Symposium, 98-105

Kotter, J. P. (1996) Leading Change. Harvard Business Press

Kwiatkowska, M. Norman, G. and Parker, D. (2006). Symmetry reduction for probabilistic model checking, in Proc. 18th International Conference on Computer Aided Verification (CAV'06)

Kwiatkowska, M., Norman, G. & Parker, D. (2011). PRISM 4.0: Verification of Probabilistic Real-time Systems. In Proc. 23rd International Conference on Computer Aided Verification (CAV'11), vol. 6806 of LNCS, 585-591, Springer

Kwiatkowska, M., Parker, D. (2012), Advances in probabilistic model checking', in Software Safety and Security - Tools for Analysis and Verification, T. Nipkow, O. Grumberg and B. Hauptmann, Eds., ser. NATO Science for Peace and Security Series - D: Information and Communication Security, IOS Press, vol. 33: 126-151

Muehlen, M., Recker, J. (2008). How Much Language Is Enough? Theoretical and Practical Use of the Business Process Modeling Notation. Proceedings for CAiSE '08 Proceedings of the 20th international conference on Advanced Information Systems Engineering, 465-479

Object Management Group (2011) Business process model and notation (BPMN) 2.0. Standards Document formal/2011-01-03. Webpage: http://www.omg.org/spec/BPMN/2.0/

Parker, D. (2012). PRISM 4.0 manual, avaliable: http://www.prismmodelchecker.org/mamual

Shingo, S. (1981) Study of Toyota Production System from Industrial Engineering Viewpoint. Japan Management Association

Vergidis, K., Tiwari, A. & Majeed, B. (2008) Business process analysis and
optimization: beyond reengineering., *IEEE Transactions on Systems, Man, and Cybernetics, Part C*, 38(1), 69–82.

White, D. J. (1993), Markov decision processes. John Wiley & Sons

Wijs, A. J., Pol, J. C. van de & Bortnik, E. M. (2009). Solving scheduling problems by untimed model checking : the clinical chemical analyser case study. International Journal on Software Tools for Technology Transfer, 11(5), 375-392

Index of Contributors

M. Schabacker et al. (eds.), *Modelling and Management of Engineering Processes*, DOI 10.1007/978-3-662-44009-4

MIX
Papier aus verantwortungsvollen Quellen
Paper from responsible sources
FSC® C105338

FSC
www.fsc.org

If you have any concerns about our products,
you can contact us on
ProductSafety@springernature.com

In case Publisher is established outside the EU,
the EU authorized representative is:
Springer Nature Customer Service Center GmbH
Europaplatz 3, 69115 Heidelberg, Germany

Printed by Libri Plureos GmbH
in Hamburg, Germany